NEW BOOK
NO REFUND IF REMOVED

A 89.95

DISCARDED BY
SUFFOLK UNIVERSITY
SAWYER LIBRARY

SUFFOLK UNIVERSITY
MILDRED F. SAWYER LIBRARY
8 ASHBURTON PLACE
BOSTON, MA 02108

Kernel Smoothing

MONOGRAPHS ON STATISTICS AND APPLIED PROBABILITY

General Editors

V. Isham, N. Keiding, T. Louis, N. Reid, R. Tibshirani, and H. Tong

1 Stochastic Population Models in Ecology and Epidemiology *M S Barlett* (1960)
2 Queues *D R Cox and W L Smith* (1961)
3 Monte Carlo Methods *J M Hammersley and D C Handscomb* (1964)
4 The Statistical Analysis of Series of Events *D R Cox and P.A W Lewis* (1966)
5 Population Genetics *W J Ewens* (1969)
6 Probability, Statistics and Time *M S Barlett* (1975)
7 Statistical Inference *S D Silvey* (1975)
8 The Analysis of Contingency Tables *B S Everitt* (1977)
9 Multivariate Analysis in Behavioural Research *A E Maxwell* (1977)
10 Stochastic Abundance Models *S Engen* (1978)
11 Some Basic Theory for Statistical Inference *E J G Pitman* (1979)
12 Point Processes *D R Cox and V Isham* (1980)
13 Identification of Outliers *D M Hawkins* (1980)
14 Optimal Design *S D Silvey* (1980)
15 Finite Mixture Distributions *B S Everitt and D J Hand* (1981)
16 Classification *A D Gordon* (1981)
17 Distribution-Free Statistical Methods, 2nd edition *J S Maritz* (1995)
18 Residuals and Influence in Regression *R D Cook and S Weisberg* (1982)
19 Applications of Queueing Theory, 2nd edition *G F Newell* (1982)
20 Risk Theory, 3rd edition *R E Beard, T Pentikäinen and E Pesonen* (1984)
21 Analysis of Survival Data *D R Cox and D Oakes* (1984)
22 An Introduction to Latent Variable Models *B S Everitt* (1984)
23 Bandit Problems *D A Berry and B Fristedt* (1985)
24 Stochastic Modelling and Control *M H A Davis and R Vinter* (1985)
25 The Statistical Analysis of Composition Data *J Aitchison* (1986)
26 Density Estimation for Statistics and Data Analysis *B W Silverman* (1986)
27 Regression Analysis with Applications *G B Wetherill* (1986)
28 Sequential Methods in Statistics, 3rd edition
G B Wetherill and K D Glazebrook (1986)
29 Tensor Methods in Statistics *P. McCullagh* (1987)
30 Transformation and Weighting in Regression
R J Carroll and D Ruppert (1988)
31 Asymptotic Techniques for Use in Statistics
O E Bandorff-Nielsen and D R Cox (1989)
32 Analysis of Binary Data, 2nd edition *D R Cox and E J Snell* (1989)
33 Analysis of Infectious Disease Data *N G Becker* (1989)
34 Design and Analysis of Cross-Over Trials *B Jones and M G Kenward* (1989)

35 Empirical Bayes Methods, 2nd edition *J S Maritz and T Lwin* (1989)
36 Symmetric Multivariate and Related Distributions
   K T Fang, S Kotz and K W Ng (1990)
37 Generalized Linear Models, 2nd edition *P. McCullagh and J A Nelder* (1989)
38 Cyclic and Computer Generated Designs, 2nd edition
   J A John and E R Williams (1995)
39 Analog Estimation Methods in Econometrics *C F Manski* (1988)
40 Subset Selection in Regression *A J Miller* (1990)
41 Analysis of Repeated Measures *M J Crowder and D J Hand* (1990)
42 Statistical Reasoning with Imprecise Probabilities *P. Walley* (1991)
43 Generalized Additive Models *T J Hastie and R J Tibshirani* (1990)
44 Inspection Errors for Attributes in Quality Control
   N L Johnson, S Kotz and X Wu (1991)
45 The Analysis of Contingency Tables, 2nd edition *B S Everitt* (1992)
46 The Analysis of Quantal Response Data *B J T Morgan* (1992)
47 Longitudinal Data with Serial Correlation—A state-space approach
   R H Jones (1993)
48 Differential Geometry and Statistics *M K Murray and J W Rice* (1993)
49 Markov Models and Optimization *M H A Davis* (1993)
50 Networks and Chaos—Statistical and probabilistic aspects
   O E Barndorff-Nielsen, J L Jensen and W S Kendall (1993)
51 Number-Theoretic Methods in Statistics *K -T Fang and Y Wang* (1994)
52 Inference and Asymptotics *O E Barndorff-Nielsen and D R Cox* (1994)
53 Practical Risk Theory for Actuaries
   C D Daykin, T Pentikäinen and M Pesonen (1994)
54 Biplots *J C Gower and D J Hand* (1996)
55 Predictive Inference—An introduction *S Geisser* (1993)
56 Model-Free Curve Estimation *M E Tarter and M D Lock* (1993)
57 An Introduction to the Bootstrap *B Efron and R J Tibshirani* (1993)
58 Nonparametric Regression and Generalized Linear Models
   P.J Green and B W Silverman (1994)
59 Multidimensional Scaling *T F Cox and M A A Cox* (1994)
60 Kernel Smoothing *M P. Wand and M C Jones* (1995)
61 Statistics for Long Memory Processes *J Beran* (1995)
62 Nonlinear Models for Repeated Measurement Data
   M Davidian and D M Giltinan (1995)
63 Measurement Error in Nonlinear Models
   R J Carroll, D Rupert and L A Stefanski (1995)
64 Analyzing and Modeling Rank Data *J J Marden* (1995)
65 Time Series Models—In econometrics, finance and other fields
   D R Cox, D V Hinkley and O E Barndorff-Nielsen (1996)
66 Local Polynomial Modeling and its Applications *J Fan and I Gijbels* (1996)
67 Multivariate Dependencies—Models, analysis and interpretation
   D R Cox and N Wermuth (1996)

68 Statistical Inference—Based on the likelihood *A Azzalini* (1996)
69 Bayes and Empirical Bayes Methods for Data Analysis
B P. Carlin and T A Louis (1996)
70 Hidden Markov and Other Models for Discrete-Valued Time Series
I L Macdonald and W Zucchini (1997)
71 Statistical Evidence—A likelihood paradigm *R Royall* (1997)
72 Analysis of Incomplete Multivariate Data *J L Schafer* (1997)
73 Multivariate Models and Dependence Concepts *H Joe* (1997)
74 Theory of Sample Surveys *M E Thompson* (1997)
75 Retrial Queues *G Falin and J G C Templeton* (1997)
76 Theory of Dispersion Models *B Jørgensen* (1997)
77 Mixed Poisson Processes *J Grandell* (1997)
78 Variance Components Estimation—Mixed models, methodologies and applications
P.S R S Rao (1997)
79 Bayesian Methods for Finite Population Sampling
G Meeden and M Ghosh (1997)
80 Stochastic Geometry—Likelihood and computation
O E Barndorff-Nielsen, W S Kendall and M N M van Lieshout (1998)
81 Computer-Assisted Analysis of Mixtures and Applications—
Meta-analysis, disease mapping and others *D Böhning* (1999)
82 Classification, 2nd edition *A D Gordon* (1999)
83 Semimartingales and their Statistical Inference *B L S Prakasa Rao* (1999)
84 Statistical Aspects of BSE and vCJD—Models for Epidemics
C A Donnelly and N M Ferguson (1999)
85 Set-Indexed Martingales *G Ivanoff and E Merzbach* (2000)
86 The Theory of the Design of Experiments *D R Cox and N Reid* (2000)
87 Complex Stochastic Systems
O E Barndorff-Nielsen, D R Cox and C Klüppelberg (2001)
88 Multidimensional Scaling, 2nd edition *T F Cox and M A A Cox* (2001)
89 Algebraic Statistics—Computational Commutative Algebra in Statistics
G Pistone, E Riccomagno and H P. Wynn (2001)
90 Analysis of Time Series Structure—SSA and Related Techniques
N Golyandina, V Nekrutkin and A A Zhigljavsky (2001)
91 Subjective Probability Models for Lifetimes
Fabio Spizzichino (2001)
92 Empirical Likelihood *Art B Owen (2001)*
93 Statistics in the 21st Century *Adrian E Raftery, Martin A Tanner,
and Martin T Wells (2001)*
94 Accelerated Life Models: Modeling and Statistical Analysis
Vilijandas Bagdonavičius and Mikhail Nikulin (2001)
95 Subset Selection in Regression, Second Edition
Alan Miller (2002)
96 Topics in Modelling of Clustered Data
Marc Aerts, Helena Geys, Geert Molenberghs, and Louise M Ryan (2002)
97 Components of Variance *D R Cox and P.J Solomon* (2002)

Kernel Smoothing

M.P. Wand
Department of Biostatistics
Harvard School of Public Health
Boston, MA, US

M.C. Jones
Department of Statistics
The Open University
Milton Keynes, UK

CHAPMAN & HALL/CRC
Boca Raton   London   New York   Washington, D.C.

Library of Congress Cataloging-in-Publication Data

Catalog record is available from the Library of Congress

This book contains information obtained from authentic and highly regarded sources Reprinted material is quoted with permission, and sources are indicated A wide variety of references are listed Reasonable efforts have been made to publish reliable data and information, but the author and the publisher cannot assume responsibility for the validity of all materials or for the consequences of their use

Neither this book nor any part may be reproduced or transmitted in any form or by any means, electronic or mechanical, including photocopying, microfilming, and recording, or by any information storage or retrieval system, without prior permission in writing from the publisher

The consent of CRC Press LLC does not extend to copying for general distribution, for promotion, for creating new works, or for resale Specific permission must be obtained in writing from CRC Press LLC for such copying

Direct all inquiries to CRC Press LLC, 2000 N W Corporate Blvd , Boca Raton, Florida 33431

Trademark Notice: Product or corporate names may be trademarks or registered trademarks, and are used only for identification and explanation, without intent to infringe

Visit the CRC Press Web site at www.crcpress.com

First edition 1995
© 1995 by M P Wand and M C Jones

No claim to original U S Government works
International Standard Book Number 0-412-55270-1
Printed in the United States of America   4 5 6 7 8 9 0
Printed on acid-free paper

Contents

Preface   xi

1 Introduction   1
    1.1 Introduction   1
    1.2 Density estimation and histograms   5
    1.3 About this book   7
    1.4 Options for reading this book   9
    1.5 Bibliographical notes   9

2 Univariate kernel density estimation   10
    2.1 Introduction   10
    2.2 The univariate kernel density estimator   11
    2.3 The MSE and MISE criteria   14
    2.4 Order and asymptotic notation; Taylor expansion   17
        2.4.1 Order and asymptotic notation   17
        2.4.2 Taylor expansion   19
    2.5 Asymptotic MSE and MISE approximations   19
    2.6 Exact MISE calculations   24
    2.7 Canonical kernels and optimal kernel theory   28
    2.8 Higher-order kernels   32
    2.9 Measuring how difficult a density is to estimate   36
    2.10 Modifications of the kernel density estimator   40
        2.10.1 Local kernel density estimators   40
        2.10.2 Variable kernel density estimators   42
        2.10.3 Transformation kernel density estimators   43
    2.11 Density estimation at boundaries   46
    2.12 Density derivative estimation   49
    2.13 Bibliographical notes   50
    2.14 Exercises   52

3 Bandwidth selection   58
    3.1 Introduction   58
    3.2 Quick and simple bandwidth selectors   59
        3.2.1 Normal scale rules   60

		3.2.2. Oversmoothed bandwidth selection rules	61
	3.3	Least squares cross-validation	63
	3.4	Biased cross-validation	65
	3.5	Estimation of density functionals	67
	3.6	Plug-in bandwidth selection	71
		3.6.1 Direct plug-in rules	71
		3.6.2 Solve-the-equation rules	74
	3.7	Smoothed cross-validation bandwidth selection	75
	3.8	Comparison of bandwidth selectors	79
		3.8.1 Theoretical performance	79
		3.8.2 Practical advice	85
	3.9	Bibliographical notes	86
	3.10	Exercises	88

4 Multivariate kernel density estimation — 90
	4.1	Introduction	90
	4.2	The multivariate kernel density estimator	91
	4.3	Asymptotic MISE approximations	94
	4.4	Exact MISE calculations	101
	4.5	Choice of a multivariate kernel	103
	4.6	Choice of smoothing parametrisation	105
	4.7	Bandwidth selection	108
	4.8	Bibliographical notes	110
	4.9	Exercises	110

5 Kernel regression — 114
	5.1	Introduction	114
	5.2	Local polynomial kernel estimators	116
	5.3	Asymptotic MSE approximations: linear case	120
		5.3.1 Fixed equally spaced design	120
		5.3.2 Random design	123
	5.4	Asymptotic MSE approximations: general case	125
	5.5	Behaviour near the boundary	126
	5.6	Comparison with other kernel estimators	130
		5.6.1 Asymptotic comparison	130
		5.6.2 Effective kernels	133
	5.7	Derivative estimation	135
	5.8	Bandwidth selection	138
	5.9	Multivariate nonparametric regression	140
	5.10	Bibliographical notes	141
	5.11	Exercises	143

6 Selected extra topics — 146
	6.1	Introduction	146
	6.2	Kernel density estimation in other settings	147

	6.2.1	Dependent data	147
	6.2.2	Length biased data	150
	6.2.3	Right-censored data	154
	6.2.4	Data measured with error	156
6.3	Hazard function estimation		160
6.4	Spectral density estimation		162
6.5	Likelihood-based regression models		164
6.6	Intensity function estimation		167
6.7	Bibliographical notes		169
6.8	Exercises		170

Appendices   **172**

A	Notation		172
B	Tables		175
C	Facts about normal densities		177
	C.1	Univariate normal densities	177
	C.2	Multivariate normal densities	180
	C.3	Bibliographical notes	181
D	Computation of kernel estimators		182
	D.1	Introduction	182
	D.2	The binned kernel density estimator	183
	D.3	Computation of kernel functional estimates	188
	D.4	Computation of kernel regression estimates	189
	D.5	Extension to multivariate kernel smoothing	191
	D.6	Computing practicalities	192
	D.7	Bibliographical notes	192

References   193

Index   208

TO

CHRISTINE, PAUL AND HANDAN

BETTY, FRANK AND PING

Preface

Kernel smoothing refers to a general class of techniques for nonparametric estimation of functions. Suppose that you have a univariate set of data which you want to display graphically. Then kernel smoothing provides an attractive procedure for achieving this goal, known as kernel density estimation. Another fundamental example is the simple nonparametric regression or scatterplot smoothing problem where kernel smoothing offers a way of estimating the regression function without the specification of a parametric model. The same principles can be extended to more complicated problems, leading to many applications in fields as diverse as medicine, engineering and economics. The simplicity of kernel estimators entails mathematical tractability, so one can delve deeply into the properties of these estimators without highly sophisticated mathematics. In summary, kernel smoothing provides simple, reliable and useful answers to a wide range of important problems.

The main goals of this book are to develop the reader's intuition and mathematical skills required for a comprehensive understanding of kernel smoothing, and hence smoothing problems in general. Exercises designed for achieving this goal have been included at the end of each chapter. We have aimed this book at newcomers to the field. These may include students and researchers from both the statistical sciences and interface disciplines. Our feeling is that this book would be appropriate for most first or second year statistics graduate students in the North American system, honours level students in the Commonwealth system and students at similar stages in other systems. In its role as an introductory text this book does make some sacrifices. It does not completely cover the vast amount of research in the field of kernel smoothing, and virtually ignores important work on non-kernel approaches to smoothing problems. It is hoped that the bibliographical notes near the end of each chapter will provide sufficient access to the wider field.

PREFACE

In completing this book we would like to extend our most sincere gratitude to Peter Hall and Steve Marron. Peter is responsible for a substantial portion of the deep theoretical understanding of kernel smoothing that has been established in recent years, and his generosity as both supervisor and colleague has been overwhelming. Steve Marron, through many conversations and countless e-mail messages, has been an invaluable source of support and information during our research in kernel smoothing. His probing research, philosophies, insight and high standards of presentation have had a strong influence on this book. We also give special thanks to David Ruppert and Simon Sheather for their advice, ideas and support. It was largely Bernard Silverman's influence that first led to M.C.J.'s interest in this topic, for which he is most grateful. Our ideas and opinions on smoothing have also been moulded by our contact, collaboration and collegiality with many other prominent researchers in the field, including Adrian Bowman, Ray Carroll, Chris Carter, John Copas, Dennis Cox, Luc Devroye, Geoff Eagleson, Joachim Engel, Randy Eubank, Jianqing Fan, Theo Gasser, Wenceslao González-Manteiga, Wolfgang Härdle, Jeff Hart, Nancy Heckman, Nils Hjort, Iain Johnstone, Robert Kohn, Oliver Linton, Hans-Georg Müller, Jens Nielsen, Doug Nychka, Byeong Park, M. Samiuddin, Bill Schucany, David Scott, Joan Staniswalis, Jim Thompson and Tom Wehrly.

Drafts of this text have been read by Glen Barnett, Angus Chow, Inge Koch, Alun Pope, Simon Sheather, Mike Smith, Frederic Udina, Sally Wood and an anonymous reader, as well as several students who participated in a course given from this book in the Department of Statistics at the University of New South Wales. Their feedback, comments and corrections are very gratefully acknowledged. Parts of this book were written while one of us (M.P.W.) was visiting Rice University, National University of Singapore and University of British Columbia. The support of these institutions is also gratefully acknowledged. We would also like to thank Yusuf Mansuri of the Australian Graduate School of Management for the excellent computing and word processing support that he has provided throughout this project.

Finally, we must express our deepest thanks to our partners Handan and Ping and our families for their constant support and encouragement.

Kensington and Milton Keynes Matt Wand
April 1994 Chris Jones

CHAPTER 1

Introduction

1.1 Introduction

Kernel smoothing provides a simple way of finding structure in data sets without the imposition of a parametric model. One of the most fundamental settings where kernel smoothing ideas can be applied is the simple regression problem, where paired observations for each of two variables are available and one is interested in determining an appropriate functional relationship between the two variables. One of the variables, usually denoted by X, is thought of as being a *predictor* for the other variable Y, usually called the *response* variable.

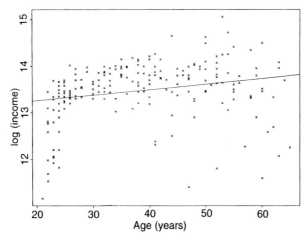

Figure 1.1. *Scatterplot of age/log(income) data. The ordinary least squares line is also shown.*

Figure 1.1 is a scatterplot with each cross representing the age

1

(the X variable) and log(income) (the Y variable) of 205 Canadian workers (source: Ullah, 1985). There is interest in modelling log(income) as a function of age. A first attempt might be to fit a straight line to the data. The line shown in Figure 1.1 is the ordinary least squares straight line fit to the observations. What are we assuming when modelling Y as a linear function of X? The usual assumption is that the sample $(X_1, Y_1), \ldots, (X_n, Y_n)$ of age/log(income) pairs satisfies the relationship

$$Y_i = \beta_0 + \beta_1 X_i + \varepsilon_i, \qquad i = 1, \ldots, n, \tag{1.1}$$

where the *errors* ε_i are symmetric random variables having zero mean. However, this assumption, which entails that the observations are randomly scattered about a straight line, appears to be far from valid for this scatterplot.

The linear model (1.1) is an example of a *parametric regression* model. Let us clarify this terminology. A well known result from elementary statistics is that the function m for which $E\{Y - m(X)\}^2$ is minimised is the conditional mean of Y given X, that is

$$m(X) = E(Y|X).$$

This function, the best mean squared predictor of Y given X, is often called the *regression* of Y on X. It follows from the definition of m that

$$Y_i = m(X_i) + \varepsilon_i, \qquad i = 1, \ldots, n$$

where $E(\varepsilon_i) = 0$ for each i. In model (1.1) we are therefore making the assumption that the functional form of the regression function m is known except for the values of the two *parameters* β_0 and β_1. This is the reason for the term *parametric* since the family of functions in the model can be specified by a finite number of parameters.

There are several other parametric regression models which one could use. Some examples are

$$Y_i = \beta_0 + \beta_1 X_i + \beta_2 X_i^2 + \varepsilon_i,$$
$$Y_i = \beta_1 \sin(\beta_2 X_i) + \varepsilon_i$$
$$\text{and} \quad Y_i = \beta_1/(\beta_2 + X_i) + \varepsilon_i.$$

The choice of parametric model depends very much on the situation. Sometimes there are scientific reasons for modelling Y as a particular function of X, while at other times the model is

1.1 INTRODUCTION

based on experience gained through analysis of previous data sets of the same type. There is, however, a drawback to parametric modelling that needs to be considered. The restriction of m belonging to a parametric family means that m can sometimes be too rigid. For example, the models above respectively require that m be parabolic, periodic or monotone, each of which might be too restrictive for adequate estimation of the true regression function. If one chooses a parametric family that is not of appropriate form, at least approximately, then there is a danger of reaching incorrect conclusions in the regression analysis.

The rigidity of parametric regression can be overcome by removing the restriction that m belong to a parametric family. This approach leads to what is commonly referred to as *nonparametric regression*. The philosophical motivation for a nonparametric approach to regression is straightforward: when confronted with a scatterplot showing no discernible simple functional form then one would want to let the data decide which function fits them best without the restrictions imposed by a parametric model (this is sometimes referred to as "letting the data speak for themselves"). However, nonparametric and parametric regression should not be viewed as mutually exclusive competitors. In many cases a nonparametric regression estimate will suggest a simple parametric model, while in other cases it will be clear that the underlying regression function is sufficiently complicated that no reasonable parametric model would be adequate.

There now exist many methods for obtaining a nonparametric regression estimate of m. Some of these are based on fairly simple ideas while others are mathematically more sophisticated. For reasons given in Section 1.3 we will study the kernel approach to nonparametric regression.

Figure 1.2 shows an estimate of m for the age/log(income) data, using what is often called a *local linear kernel estimator*. The function shown at the bottom of the plot is a *kernel* function which is usually taken to be a symmetric probability density such as a normal density. The value of the estimate at the first point u is obtained by fitting a straight line to the data using *weighted* least squares, where the weights are chosen according to the height of the kernel function. This means that the data points closer to u have more influence on the linear fit than those far from u. This local straight line fit is shown by the dotted curve and the regression estimate at u is the height of the line at u. The estimate at a different point v is found the same way, but with the weights chosen according to the heights of the kernel when centred around v. This estimator fits into the class of *local polynomial* regression estimates (Cleveland, 1979). Nonparametric regression estimators are often called *regression smoothers* or *scatterplot smoothers*, while those based on kernel functions are often called *kernel smoothers*.

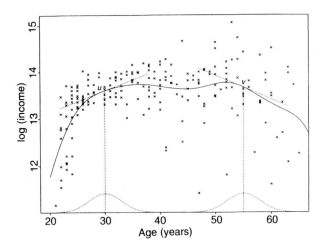

Figure 1.2. *Local linear kernel regression estimate based on the age/log(income) data. The solid curve is the estimate. The dotted curves are the kernel weights and straight line fits at points u and v.*

While the kernel-based nonparametric regression estimator described here means that we have a much more flexible family of curves to choose from, this increased flexibility has its costs and leads to several new questions. Some examples are

- What are the statistical properties of kernel regression estimators?
- What influence does the shape of the kernel function have on the estimator?
- What influence does the scaling of the kernel function have on the estimator?
- How can this scaling be chosen in practice?
- How can kernel smoothing ideas be used to make confidence statements rather than just giving point estimates?
- How do dependencies in the data affect the kernel regression estimator?
- How does one best deal with multiple predictor variables?

While some of these questions have reasonably straightforward solutions, others are the subject of ongoing research and may not

1.2 DENSITY ESTIMATION AND HISTOGRAMS

be completely resolved for many years. Some can be answered by relatively simple mathematical arguments, while others require deeper analyses that are beyond the scope of this book.

We chose to introduce the motivation and ideas of kernel smoothing by nonparametric regression because of the familiarity of the regression problem to most readers. However, as we will see, kernel smoothing can be applied to many other important curve estimation problems such as estimating probability density functions, spectral densities and hazard rate functions. The first of these is discussed in the next section.

1.2 Density estimation and histograms

Perhaps an even more fundamental problem than the regression problem is the estimation of the common probability density function, or *density* for short, of a univariate random sample. Suppose that $X_1, \ldots, X_n$ is a set of continuous random variables having common density f. The parametric approach to estimation of f involves assuming that f belongs to a parametric family of distributions, such as the normal or gamma family, and then estimating the unknown parameters using, for example, maximum likelihood estimation. On the other hand, a *nonparametric density estimator* assumes no pre-specified functional form for f.

The oldest and most widely used nonparametric density estimator is the *histogram*. This is usually formed by dividing the real line into equally sized intervals, often called *bins*. The histogram is then a step function with heights being the proportion of the sample contained in that bin divided by the width of the bin. If we let b denote the width of the bins, usually called the *binwidth*, then the histogram estimate at a point x is given by

$$\hat{f}_H(x;b) = \frac{\text{number of observations in bin containing } x}{nb}.$$

Two choices have to be made when constructing a histogram: the binwidth and the positioning of the bin edges. Each of these choices can have a significant effect on the resulting histogram. Figure 1.3 show four histograms based on the same set of data. These data represent 50 birthweights of children having severe idiopathic respiratory syndrome (source: van Vliet and Gupta, 1973). The first two histograms are based on a small and a large binwidth ($b = 0.2$ and $b = 0.8$) respectively. The bottom two are based on the same medium sized binwidth ($b = 0.4$), but with bin edges shifted by half a binwidth.

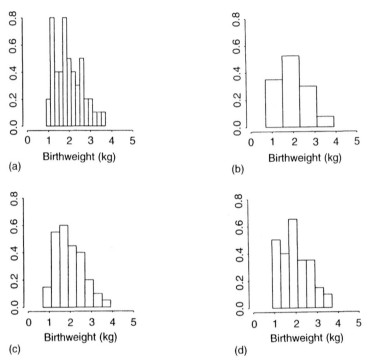

Figure 1.3. *Histograms of birthweight data. Figures (a) and (b) are based on binwidths of 0.2 and 0.8 respectively. Figures (c) and (d) are each based on a binwidth of 0.4 but with left bin edge at 0.7 and 0.9 respectively.*

Notice that each histogram gives a different impression of the shape of the density of the data. A smaller binwidth leads to a relatively jagged histogram, while a larger binwidth results in a smoother looking histogram as shown in Figures 1.3 (a) and (b). Figures 1.3 (c) and (d) show that the placement of the bin edges also has an effect since the density shapes suggested by these histograms are quite different from each other, despite the equal binwidths.

The binwidth b is usually called a *smoothing parameter* since it controls the amount of "smoothing" being applied to the data.

All nonparametric curve estimates have an associated smoothing parameter. We will see in the following chapters that, for kernel estimators, the scale of the kernel plays a role analogous to that of the binwidth. In parametric polynomial regression the degree of the polynomial can be thought of as being a smoothing parameter.

The sensitivity of the histogram to the placement of the bin edges is a problem not shared by other density estimators such as the kernel density estimator introduced in Chapter 2. The bin edge problem is one of the histogram's main disadvantages. The logical remedy to this problem is the *average shifted histogram* (Scott, 1985), which averages several histograms based on shifts of the bin edges, but this can be shown to approximate a kernel density estimator, and thus provide an appealing motivation for kernel methods. The histogram has several other problems not shared by kernel density estimators. Most densities are not step functions, yet the histogram has the unattractive feature of estimating all densities by a step function. A further problem is the extension of the histogram to the multivariate setting, especially the graphical display of a multivariate histogram. Finally, the histogram can be shown not to use the data as efficiently as the kernel estimator. This deficiency is discussed at the end of Section 2.5. Despite these drawbacks, the simplicity of histograms ensures their continuing popularity.

1.3 About this book

As the title and the previous two sections suggest, this book is about kernel smoothing as a means of obtaining nonparametric curve estimators. Kernel estimators have the advantage of being very intuitive and relatively simple to analyse mathematically. Even if one prefers to use other nonparametric smoothing methods, such as those based on spline functions, an understanding of the main issues involved can best be gained through studying kernel estimators.

Kernel estimators have been around since the seminal papers of Rosenblatt (1956) and Parzen (1962), although the basic principles were independently introduced by Fix and Hodges (1951) and Akaike (1954). Since then articles written about kernel estimators number in the thousands, and there will be many more written in the future. It follows that there are still many unresolved and controversial issues. This book makes no endeavour to survey the field of kernel estimation, nor does it try to provide an answer for every question concerning the practical implementation of kernel estimators. Instead our goal is to present the reader with the aspects of kernel smoothing which we see as being most fundamental and practically most relevant at the time of writing.

The choice of topics in the pursuit of this goal is necessarily a personal one. Moreover, because of ongoing research into the practical implementation of kernel estimators we will, in the main, avoid the more unsettled issues in the field. The main purpose of this book is to enhance the reader's intuition and mathematical skills required for understanding kernel smoothing, and hence smoothing problems in general.

We believe that the readability of the book is improved by postponing detailed referencing to bibliographical notes, which are provided near the end of each chapter. These are followed by a set of exercises which aim to familiarise the reader with the above-mentioned mathematical skills.

We begin our study of kernel smoothing with the univariate kernel density estimator in Chapter 2. This is because kernel density estimation provides a very simple and convenient way of developing an understanding of the main ideas and issues involved in kernel smoothing in general.

As we show in Chapter 2, one of the central issues in kernel smoothing is the choice of the smoothing parameter, often called the *bandwidth* for kernel estimators. The choice of the bandwidth from the data has become an important topic in its own right in recent years and is still a burgeoning area of research. In Chapter 3 we discuss some of the more popular approaches to bandwidth selection, again in the simple density estimation context. While our coverage of this topic is far from complete, our aim is to give the reader a flavour for the types of approaches and problems faced when selecting a bandwidth from the data.

Chapter 4 is devoted to the extension of the kernel density estimator to multivariate data. While this extension is fairly obvious in principle, the mathematical analyses and practical implementation are non-trivial, as Chapter 4 shows.

In Chapter 5 we return to the important problem of nonparametric regression. The notions of kernel smoothing learnt from studying the kernel density estimator prove to be useful for understanding the more complicated kernel regression problem.

Chapter 6 is a collection of extra topics from the kernel smoothing literature which portray various extensions of the material from the previous chapters, and indicate the wide applicability of the general ideas.

There are four appendices. Appendix A lists notation used throughout this book. Tables in Appendix B contain useful results for several common densities. Facts about the normal density, relevant to kernel smoothing, are given in Appendix C. Appendix D describes computation of kernel estimates.

1.4 Options for reading this book

This book may be read in several different ways. To gain a basic understanding of the main ideas of univariate kernel estimation for important settings one should consult Chapters 2 and 5. Chapter 4 could be added for an understanding of multivariate kernel smoothing. Chapter 3 stands out as the only chapter concerned with full data-driven implementation of kernel estimators, but is also the least settled topic in this book, and could be omitted without loss of continuity. Chapter 6 is a selection of extra topics which could be covered depending on time and interest. We have ordered the chapters in what we see as being the most natural sequence if this book is to be completely covered.

1.5 Bibliographical notes

For a detailed study of the histogram we refer the reader to Scott (1992).

Important references for kernel smoothing will be given at the end of the appropriate chapter. At this point we will just mention some recent books on the subject. A very readable introduction to kernel density estimation is given by Silverman (1986). Kernel density estimation is also treated in books by Devroye and Györfi (1985), Härdle (1990a) and Scott (1992). Recent books that treat nonparametric kernel regression to varying extents are Eubank (1988), Müller (1988), Härdle (1990b) and Hastie and Tibshirani (1990). Wahba (1990) and Green and Silverman (1994) are recent monographs on spline approaches to nonparametric regression, and Tarter and Lock (1993) treats orthogonal series density estimation.

CHAPTER 2

Univariate kernel density estimation

2.1 Introduction

There are two main reasons for beginning our study of kernel smoothing with the univariate kernel density estimator. The first is that nonparametric density estimation is an important data analytic tool which provides a very effective way of showing structure in a set of data at the beginning of its analysis. This was demonstrated in Section 1.2. It is especially effective when standard parametric models are inappropriate. This is illustrated in Figures 2.1 (a) and 2.1 (b) which show density estimates based on data corresponding to the incomes of 7,201 British households for the year 1975 (see e.g. Park and Marron, 1990). The data have been divided by the sample average. Figure 2.1 (a) is a parametric density estimate based on modelling income by the lognormal family of distributions which have densities of the form

$$f(x; \theta_1, \theta_2) = \phi\{(\ln x - \theta_1)/\theta_2\}/(\theta_2 x), \quad x > 0,$$

where $\phi(x) = (2\pi)^{-1/2} e^{-x^2/2}$, $-\infty < \theta_1 < \infty$ and $\theta_2 > 0$. The values of θ_1 and θ_2 were chosen by maximum likelihood. Figure 2.1 (b) is a nonparametric kernel estimate based on a version of the transformation kernel density estimator as described in Section 2.10. The main lesson is that the interesting bimodal structure (which has important economic significance) that is uncovered by the kernel estimator is completely missed by imposing a unimodal parametric model such as the lognormal. Since one of the main aims of data analysis is to highlight important structure in the data it is desirable to have a density estimator that does not assume that the density has a particular functional form.

2.2 THE UNIVARIATE KERNEL DENSITY ESTIMATOR

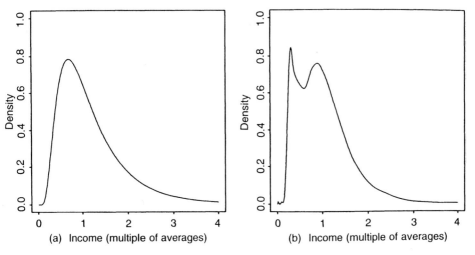

Figure 2.1. *Density estimates based on British incomes data using (a) a parametric lognormal model and (b) a transformation kernel density estimator.*

Our second reason for beginning with the density estimation setting is that the univariate kernel density estimator is the most straightforward among several types of kernel estimators. Its simplicity allows one to study its properties quite thoroughly. This makes it an excellent starting point for a newcomer to the field. Extensions from univariate to multivariate, from the basic to more sophisticated kernel-based methods, and from estimating densities to other curves, will be easier to understand given this background.

Throughout this chapter, it will be assumed that we have a random sample $X_1, \ldots, X_n$ taken from a continuous, univariate density f. At all times, an unqualified integral sign $\int$ will be taken to mean integration over the entire real line, $\mathbb{R}$. The notation $\phi_\sigma(x) = (2\pi\sigma^2)^{-1/2} \exp\{-x^2/(2\sigma^2)\}$ will be used to denote the $N(0, \sigma^2)$ density.

2.2 The univariate kernel density estimator

We will begin with the formula for the kernel density estimator,

$$\hat{f}(x; h) = (nh)^{-1} \sum_{i=1}^{n} K\{(x - X_i)/h\}. \qquad (2.1)$$

Here, K is a function satisfying $\int K(x)\,dx = 1$, which we call the *kernel*, and h is a positive number, usually called the *bandwidth* or *window width*. A slightly more compact formula for the kernel estimator can be obtained by introducing the rescaling notation $K_h(u) = h^{-1}K(u/h)$. This allows us to write

$$\hat{f}(x;h) = n^{-1}\sum_{i=1}^{n} K_h(x - X_i). \tag{2.2}$$

Usually K is chosen to be a unimodal probability density function that is symmetric about zero. This ensures that $\hat{f}(x;h)$ is itself also a density. However, kernels that are not densities are also sometimes used (see Section 2.8). Figure 2.2 shows a kernel density estimate constructed using five observations with the kernel chosen to be the $N(0,1)$ density, $K(x) = \phi(x)$. We should point out that we use just five observations here purely for clarity in illustrating how the kernel method works. Practical density estimation usually involves a much higher number of observations.

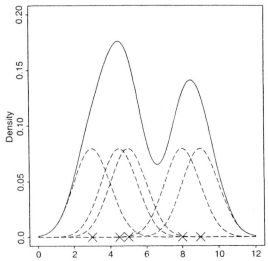

Figure 2.2. *Kernel density estimate based on five observations.*

In this case notice that K_h is simply the $N(0, h^2)$ density so that h plays the role of a scaling factor which determines the spread of the kernel. The kernel estimate is constructed by centring a scaled kernel at each observation. The value of the kernel estimate at the point x is simply the average of the n kernel ordinates at that point. One can think of the kernel as spreading a "probability mass" of size $1/n$ associated with each data point

2.2 THE UNIVARIATE KERNEL DENSITY ESTIMATOR

about its neighbourhood. Combining contributions from each data point means that in regions where there are many observations, and it is expected that the true density has a relatively large value, the kernel estimate should also assume a relatively large value. The opposite should occur in regions where there are relatively few observations.

We will see later (Section 2.7) that the choice of the shape of the kernel function is not a particularly important one. However, the choice of value for the bandwidth is very important. Figure 2.3 shows three kernel density estimates based on a sample of size $n = 1000$ from the density

$$f_1(x) = \tfrac{3}{4}\phi(x) + \tfrac{1}{4}\phi_{1/3}(x - \tfrac{3}{2}). \tag{2.3}$$

This is a normal mixture distribution which consists of $N(0, 1)$ observations with probability $\tfrac{3}{4}$ and $N(\tfrac{3}{2}, (\tfrac{1}{3})^2)$ observations with probability $\tfrac{1}{4}$. The amount of kernel smoothing done in each case is indicated by the placement of the scaled kernel K_h at the base of each graph. In Figure 2.3 (a), with $h = 0.06$, it is seen that the narrowness of the kernel means that the averaging process performed at each point is based on relatively few observations resulting in a very rough estimate of f. This estimate pays too much attention to the particular data set at hand and does not allow for the variation across samples. Such an estimate is said to be *undersmoothed*.

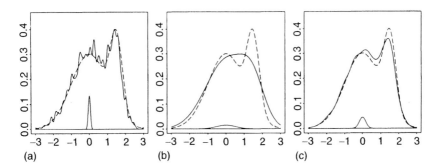

Figure 2.3. *Kernel density estimates based on a sample of $n = 1000$ observations from the normal mixture distribution f_1 described in the text. The solid line is the estimate, the broken line is the true density. The bandwidths are (a) $h = 0.06$, (b) $h = 0.54$ and (c) $h = 0.18$. The kernel weight for each estimate is illustrated by small kernels at the base of each figure.*

Figure 2.3 (b) shows another kernel estimate based on the same data, but with $h = 0.54$. This results in a much smoother estimate which is really too smooth since the bimodality structure has been smoothed away. This is an example of an *oversmoothed* estimate. In Figure 2.3 (c) a compromise is reached with $h = 0.18$. In this case, the kernel estimate is not overly noisy, yet the essential structure of the underlying density has been recovered, as comparison with the true f shows.

2.3 The MSE and MISE criteria

Our analysis of the performance of the kernel density estimator will require the specification of appropriate error criteria for measuring the error when estimating the density at a single point as well as the error when estimating the density over the whole real line.

In classical parametric statistics it is common to measure the closeness of an estimator $\hat{\theta}$ to its target parameter θ by the size of the mean squared error (MSE)

$$\text{MSE}(\hat{\theta}) = E(\hat{\theta} - \theta)^2.$$

One appealing feature of MSE is its simple decomposition into variance and squared bias

$$\text{MSE}(\hat{\theta}) = \text{Var}(\hat{\theta}) + (E\hat{\theta} - \theta)^2.$$

This error criterion is often preferred to other criteria such as mean absolute error $\text{MAE}(\hat{\theta}) = E|\hat{\theta} - \theta|$ since it is mathematically simpler to work with. The variance-bias decomposition allows easier analysis and interpretation of the performance of the kernel density estimator as we will now show.

Consider $\hat{f}(x; h)$ as an estimator of the density function $f(x)$ at some point $x \in \mathbb{R}$. To compute $\text{MSE}\{\hat{f}(x; h)\}$ we will require expressions for the mean and variance of $\hat{f}(x; h)$. These can be derived directly from (2.2). Let X be a random variable having density f. Firstly,

$$E\hat{f}(x; h) = E\, K_h(x - X) = \int K_h(x - y) f(y) \, dy. \qquad (2.4)$$

At this stage it is convenient to introduce the convolution notation,

$$(f * g)(x) = \int f(x - y) g(y) \, dy,$$

2.3 THE MSE AND MISE CRITERIA

since this allows us to write the bias of $\hat{f}(x;h)$ as

$$E\hat{f}(x;h) - f(x) = (K_h * f)(x) - f(x).$$

Convolution is usually thought of as a smoothing operation, so the bias is the difference between a smoothing of f and f itself. Similar calculations (Exercise 2.1) lead to

$$\text{Var}\{\hat{f}(x;h)\} = n^{-1}\{(K_h^2 * f)(x) - (K_h * f)^2(x)\}, \qquad (2.5)$$

and these may be combined to give

$$\begin{aligned}\text{MSE}\{\hat{f}(x;h)\} = &\, n^{-1}\{(K_h^2 * f)(x) - (K_h * f)^2(x)\} \\ &+ \{(K_h * f)(x) - f(x)\}^2.\end{aligned} \qquad (2.6)$$

Rather than simply estimating f at a fixed point, it is usually desirable, especially from a data analytic viewpoint, to estimate f over the entire real line. In this case our estimate is the function $\hat{f}(\cdot;h)$ so we need to consider an error criterion that globally measures the distance between the functions $\hat{f}(\cdot;h)$ and f. One such error criterion is the *integrated squared error* (ISE) given by

$$\text{ISE}\{\hat{f}(\cdot;h)\} = \int \{\hat{f}(x;h) - f(x)\}^2 \, dx.$$

This may be recognized as the square of the L_2 distance between $\hat{f}(\cdot;h)$ and f. The ISE is appropriate if we are only concerned with the data set at hand, but it does not take into account other possible data sets from the density f. Therefore, it will be more appropriate to analyse the expected value of this random quantity, the *mean integrated squared error* (MISE),

$$\text{MISE}\{\hat{f}(\cdot;h)\} = E[\text{ISE}\{\hat{f}(\cdot;h)\}] = E\int \{\hat{f}(x;h) - f(x)\}^2 \, dx.$$

Note that by changing the order of integration we have

$$\begin{aligned}\text{MISE}\{\hat{f}(\cdot;h)\} &= \int E\{\hat{f}(x;h) - f(x)\}^2 \, dx \\ &= \int \text{MSE}\{\hat{f}(x;h)\} \, dx\end{aligned}$$

so we can use (2.6) to obtain

$$\mathrm{MISE}\{\hat{f}(\cdot;h)\} = n^{-1} \int \{(K_h^2 * f)(x) - (K_h * f)^2(x)\}\, dx$$
$$+ \int \{(K_h * f)(x) - f(x)\}^2\, dx. \qquad (2.7)$$

However, some straightforward manipulations (Exercise 2.2) lead to the more manageable expression

$$\mathrm{MISE}\{\hat{f}(\cdot;h)\} = (nh)^{-1} \int K^2(x)\, dx$$
$$+ (1 - n^{-1}) \int (K_h * f)^2(x)\, dx \qquad (2.8)$$
$$- 2 \int (K_h * f)(x) f(x)\, dx + \int f(x)^2\, dx.$$

While this is a relatively compact expression for $\mathrm{MISE}\{\hat{f}(\cdot;h)\}$ it has the misfortune of depending on the bandwidth h in a fairly complicated fashion. In Section 2.5 we will derive an asymptotic approximation to (2.6) which depends on h in a very simple way and allows much greater appreciation of the effect of this parameter. In Section 2.6 we will look at some situations where this expression can be further simplified.

We should stress that our decision to work with MISE for measuring the global performance of the kernel density estimator is largely because of its mathematical simplicity. There are good reasons for working with other criteria such as the *mean integrated absolute error* (MIAE) given by

$$\mathrm{MIAE}\{\hat{f}(\cdot;h)\} = E \int |\hat{f}(x;h) - f(x)|\, dx.$$

These include the fact that MIAE is always defined when $\hat{f}(\cdot;h)$ is a density, and invariance of MIAE under monotone transformations (see Devroye and Györfi, 1985, p.1). However, the analysis of this quantity is substantially more complicated.

It should also be realised that MISE and MIAE do not necessarily conform with human perception of closeness of a density estimate to its target. This has motivated the development of new error criteria that are more satisfactory in this respect (Marron and Tsybakov, 1993).

2.4 Order and asymptotic notation; Taylor expansion

The purpose of this section is to provide the uninitiated reader with the main mathematical tools used for the large sample analysis of the kernel density estimator to follow. Otherwise, this section is completely separate from the density estimation problem.

2.4.1 Order and asymptotic notation

The order notation O and o is defined for general real-valued functions (e.g. Serfling, 1980, p.1). However, we will only require the notation for real-valued sequences and so will restrict our attention to this case.

Let a_n and b_n each be sequences of real numbers. Then we will say that a_n is of order b_n (or a_n is "big oh" b_n) as $n \to \infty$, and write

$$a_n = O(b_n) \quad \text{as } n \to \infty, \text{ if and only if } \limsup_{n \to \infty} |a_n/b_n| < \infty.$$

In other words, $a_n = O(b_n)$ if $|a_n/b_n|$ remains bounded as $n \to \infty$. We say that a_n is of small order b_n (or a_n is "small oh" b_n), and write

$$a_n = o(b_n) \quad \text{as } n \to \infty, \text{ if and only if } \lim_{n \to \infty} |a_n/b_n| = 0.$$

Since, in the case of sequences, it is usually understood that n approaches ∞, the "as $n \to \infty$" will be taken as given.

EXAMPLE. If $c_n = (n+4)/(n^3+1)$ then we may write $c_n = O(n^{-2})$. It should be noted that the order of a sequence is not unique. We could also write $c_n = O(n^{-1})$ or even $c_n = O(n^2)$ and the definition would still be satisfied.

The small order of a sequence is also not unique since it is valid to write $c_n = o(n^{-1})$ and $c_n = o(1)$, for example. ∎

Note that $a_n = O(1)$ is equivalent to the sequence a_n being bounded and that $a_n = o(1)$ is equivalent to $a_n \to 0$ as $n \to \infty$.

Also we say that a_n is *asymptotically equivalent to* b_n, or simply a_n is *asymptotic to* b_n, and write

$$a_n \sim b_n, \quad \text{if and only if } \lim_{n \to \infty}(a_n/b_n) = 1.$$

EXAMPLE. It follows from the power series expansion of the cosine

function that
$$1 - \cos\{(2n^{1/2} + 3)/(3n^2 + 1)\} \sim \tfrac{2}{9}n^{-3}.$$

■

It is usual to express the right hand side of $\sim$ in the form Cr_n where r_n is a simple function of n such as one of the form $n^{-\alpha}$ or $(\log n)^{-\alpha}$ and C is independent of n. If the sequence a_n satisfies $a_n \sim Cr_n$ then we usually call r_n the *rate of convergence* to zero of a_n or say that a_n has *leading term* Cr_n. It is also common to say that a_n *is of order* r_n. We often refer to C as the *constant coefficient* or simply the *constant*.

In estimation theory one is often able to obtain the rate of convergence of some non-random error criterion, such as MSE, as the sample size n increases, even in situations where the exact error cannot be formulated explicitly. The concept of rate of convergence has the simple interpretation of showing how "quickly" an estimator approaches its target as the sample size grows and can be very useful for comparison of different competing estimators.

EXAMPLE. Suppose that we have a random sample $X_1, \ldots, X_n$ from the $N(\mu, \sigma^2)$ distribution and the parameter of interest is $\exp(\mu)$. The maximum likelihood estimator of this parameter is $\exp(\overline{X})$ where $\overline{X}$ is the sample mean. It can be shown (Exercise 2.5) that

$$\text{MSE}\{\exp(\overline{X})\} = e^{2\mu}(e^{2\sigma^2/n} - 2e^{\sigma^2/(2n)} + 1). \qquad (2.9)$$

While this is an explicit expression for the MSE it is difficult to appreciate the effect of increasing the sample size. However, from the power series expansion of the exponential function we have

$$\text{MSE}\{\exp(\overline{X})\} = e^{2\mu}(\sigma^2 n^{-1} + \tfrac{7}{4}\sigma^4 n^{-2} + \ldots)$$

which leads to
$$\text{MSE}\{\exp(\overline{X})\} \sim \sigma^2 e^{2\mu} n^{-1}.$$

This expression provides a good approximation to the error which is much easier to comprehend than (2.9).

■

In the preceding example the model is parametric since the only unknown parameters are μ and σ^2. Notice that the MSE has a rate of convergence of order n^{-1}. This rate is typical for MSE in parametric settings. In our study of nonparametric kernel estimators we will see that rates of convergence are typically slower than n^{-1}.

2.4.2 Taylor expansion

A vital mathematical tool for obtaining asymptotic approximations in kernel smoothing is Taylor expansion. This relies on a version of Taylor's theorem which we now state.

TAYLOR'S THEOREM. *Suppose that f is a real-valued function defined on $\mathbb{R}$ and let $x \in \mathbb{R}$. Assume that f has p continuous derivatives in an interval $(x - \delta, x + \delta)$ for some $\delta > 0$. Then for any sequence α_n converging to zero,*

$$f(x + \alpha_n) = \sum_{j=0}^{p} (\alpha_n^j/j!) f^{(j)}(x) + o(\alpha_n^p).$$

Taylor's theorem allows us to approximate function values close to a given point in terms of higher-order derivatives at that point, provided the function is smooth enough so that these derivatives exist.

EXAMPLE. If α_n is a sequence converging to zero then Taylor's theorem gives

$$(x + \alpha_n)^{-1/2} = x^{-1/2} - \tfrac{1}{2}\alpha_n x^{-3/2} + \tfrac{3}{8}\alpha_n^2 x^{-5/2} + o(\alpha_n^2)$$

for all $x > 0$.

∎

2.5 Asymptotic MSE and MISE approximations

A problem with the MSE and MISE expressions (2.6) and (2.8) is that they depend on the bandwidth in a complicated way. This makes it difficult to interpret the influence of the bandwidth on the performance of the kernel density estimator. In this section we investigate one way of overcoming this problem that involves the derivation of large sample approximations for leading variance and bias terms. These approximations have very simple expressions that allow a deeper appreciation of the role of the bandwidth. They can also be used to obtain the rate of convergence of the kernel density estimator and the MISE-optimal bandwidth.

Throughout this section we will make the following assumptions.

(i) The density f is such that its second derivative f'' is continuous, square integrable and ultimately monotone.

(ii) The bandwidth $h = h_n$ is a non-random sequence of positive numbers. To keep the notation less cumbersome the dependence of h on n will be suppressed in our calculations. We also assume that h satisfies

$$\lim_{n \to \infty} h = 0 \quad \text{and} \quad \lim_{n \to \infty} nh = \infty,$$

which is equivalent to saying that h approaches zero, but at a rate slower than n^{-1}.

(iii) The kernel K is a bounded probability density function having finite fourth moment and symmetry about the origin.

An ultimately monotone function is one that is monotone over both $(-\infty, -M)$ and (M, ∞) for some $M > 0$. Note that conditions (i) and (iii) can be replaced by numerous other combinations of conditions on f and K so that the results presented here remain valid.

We first consider the estimation of $f(x)$ at $x \in \mathbb{R}$. We have, from (2.4) and a change of variables,

$$E\hat{f}(x; h) = \int K(z) f(x - hz) \, dz.$$

Expanding $f(x - hz)$ in a Taylor series about x we obtain

$$f(x - hz) = f(x) - hzf'(x) + \tfrac{1}{2} h^2 z^2 f''(x) + o(h^2)$$

uniformly in z. This leads to

$$E\hat{f}(x; h) = f(x) + \tfrac{1}{2} h^2 f''(x) \int z^2 K(z) \, dz + o(h^2)$$

where we have used

$$\int K(z) \, dz = 1, \quad \int z K(z) \, dz = 0 \quad \text{and} \quad \int z^2 K(z) \, dz < \infty,$$

each of which follow from assumption (iii). We now introduce the notation $\mu_2(K) = \int z^2 K(z) \, dz$, which leads to the bias expression

$$E\hat{f}(x; h) - f(x) = \tfrac{1}{2} h^2 \mu_2(K) f''(x) + o(h^2). \tag{2.10}$$

Notice that the bias is of order h^2 which implies that $\hat{f}(x; h)$ is asymptotically unbiased. Also noteworthy is the way that the bias

2.5 ASYMPTOTIC MSE AND MISE APPROXIMATIONS

depends on the true f. The bias is large whenever the absolute value of the second derivative of f is large, and this happens in regions where the curvature of the density is high. For many densities this occurs in peaks where the bias is negative, and valleys where the bias is positive, which shows that $\hat{f}(x;h)$ has a tendency to smooth out such features on average.

For the variance, note that from (2.5),

$$\begin{aligned}\operatorname{Var}\{\hat{f}(x;h)\} &= (nh)^{-1}\int K(z)^2 f(x-hz)\,dz \\ &\quad - n^{-1}\left\{E\hat{f}(x;h)\right\}^2 \\ &= (nh)^{-1}\int K(z)^2\{f(x)+o(1)\}\,dz \\ &\quad - n^{-1}\{f(x)+o(1)\}^2 \\ &= (nh)^{-1}\int K(z)^2\,dz\, f(x) + o\{(nh)^{-1}\}.\end{aligned}$$

Another useful notation is $R(g) = \int g(x)^2\,dx$ for any square-integrable function g. This allows us to write the variance as

$$\operatorname{Var}\{\hat{f}(x;h)\} = (nh)^{-1}R(K)f(x) + o\{(nh)^{-1}\}. \qquad (2.11)$$

Since the variance is of order $(nh)^{-1}$ assumption (ii) ensures that $\operatorname{Var}\{\hat{f}(x;h)\}$ converges to zero.

Adding (2.11) and the square of (2.10) we obtain

$$\begin{aligned}\operatorname{MSE}\{\hat{f}(x;h)\} &= (nh)^{-1}R(K)f(x) + \tfrac{1}{4}h^4\mu_2(K)^2 f''(x)^2 \\ &\quad + o\{(nh)^{-1}+h^4\}.\end{aligned}$$

If we integrate this expression then, under our integrability assumption on f, we obtain

$$\operatorname{MISE}\{\hat{f}(\cdot;h)\} = \operatorname{AMISE}\{\hat{f}(\cdot;h)\} + o\{(nh)^{-1}+h^4\}$$

where

$$\operatorname{AMISE}\{\hat{f}(\cdot;h)\} = (nh)^{-1}R(K) + \tfrac{1}{4}h^4\mu_2(K)^2 R(f''). \qquad (2.12)$$

We call this the *asymptotic* MISE since it provides a useful large sample approximation to the MISE. The AMISE is a much simpler expression to comprehend than the expression for the MISE given by (2.8). Notice that the integrated squared bias is asymptotically proportional to h^4, so for this quantity to decrease one needs to take h to be small. However, taking h small means an increase in the leading term of the integrated variance since this quantity is proportional to $(nh)^{-1}$. Therefore, as n increases h should vary in such a way that each of the components of the MISE becomes smaller. This is known as the *variance-bias trade-off* and is a mathematical quantification for the critical role of the bandwidth. This variance-bias trade-off is in accordance with the intuitive role of h discussed in Section 2.2 via Figure 2.3. For very small h, $\hat{f}(\cdot;h)$ is very spiky and hence very variable in the sense that, over repeated sampling from f, the spikes would appear in different places. There is, however, very little bias. If more smoothing is performed, that is h is increased, then the variability is reduced at the expense of introducing bias: for increasingly large h, there would be large bias because all features are eventually smoothed away, but little variance because the data are essentially ignored.

Another advantage of AMISE is that the optimal bandwidth with respect to this criterion, which we denote by h_{AMISE}, has a closed form expression,

$$h_{\text{AMISE}} = \left[\frac{R(K)}{\mu_2(K)^2 R(f'')n}\right]^{1/5}. \qquad (2.13)$$

This can be easily derived by differentiating (2.12) with respect to h and setting the derivative equal to zero. Aside from its dependence on the known K and n, this expression shows us that h_{AMISE} is inversely proportional to $R(f'')^{1/5}$. Since $|f''(x)|$ is a measure of the curvature of f, the functional $R(f'')$ measures the total curvature of f. Thus for a density with little curvature, $R(f'')$ will be small and a large bandwidth is called for; when $R(f'')$ is large, on the other hand, little smoothing will be optimal. Unfortunately, direct use of (2.13) to choose a good bandwidth in practice is not possible since $R(f'')$ is unknown. However, there now exist several rules for selecting h based on estimating $R(f'')$. Some of these proposals are discussed in Chapter 3.

Substituting (2.13) into (2.12) leads to

$$\inf_{h>0} \text{AMISE}\{\hat{f}(\cdot;h)\} = \tfrac{5}{4}\{\mu_2(K)^2 R(K)^4 R(f'')\}^{1/5} n^{-4/5} \qquad (2.14)$$

which is the smallest possible AMISE for estimation of f using the kernel K.

2.5 ASYMPTOTIC MSE AND MISE APPROXIMATIONS

We can rewrite the information conveyed by (2.13) and (2.14) in terms of the MISE itself using the asymptotic notation. Thus, if h_{MISE} is the minimiser of $\text{MISE}\{\hat{f}(\cdot;h)\}$ then

$$h_{\text{MISE}} \sim \left[\frac{R(K)}{\mu_2(K)^2 R(f'')n}\right]^{1/5} \quad \text{and}$$

$$\inf_{h>0} \text{MISE}\{\hat{f}(\cdot;h)\} \sim \tfrac{5}{4}\{\mu_2(K)^2 R(K)^4 R(f'')\}^{1/5} n^{-4/5}.$$

These expressions give the rate of convergence of the MISE-optimal bandwidth and the minimum MISE, respectively, to zero as $n \to \infty$. Under the stated assumptions, the best obtainable rate of convergence of the MISE of the kernel estimator is of order $n^{-4/5}$. As mentioned in Section 2.4, MSE rates slower than the typical parametric rate of order n^{-1} are to be expected in nonparametric function estimation.

Asymptotic MISE approximations can also be used to make comparisons of the kernel estimator to the histogram. Recall from Section 1.2 that b is the binwidth of the histogram $\hat{f}_H(\cdot;b)$. If b satisfies assumption (ii) and f has a continuous square integrable first derivative then it can be shown that

$$\text{AMISE}\{\hat{f}_H(\cdot;b)\} = (nb)^{-1} + \tfrac{1}{12} b^2 R(f')$$

(Scott, 1979). Notice in particular that the integrated squared bias of the histogram is of order b^2 which is considerably larger than the $O(h^4)$ integrated squared bias of the kernel estimator. It follows that

$$b_{\text{MISE}} \sim \{6/R(f')\}^{1/3} n^{-1/3}$$

$$\text{and} \quad \inf_{b>0} \text{MISE}\{\hat{f}_H(\cdot;b)\} \sim \tfrac{1}{4}\{36 R(f')\}^{1/3} n^{-2/3}$$

(Scott, 1979). Therefore, the MISE of the histogram is asymptotically inferior to the kernel density estimator since its convergence rate is $O(n^{-2/3})$ compared to the kernel estimator's $O(n^{-4/5})$ rate. This is a mathematical quantification of the inefficiency of the histogram mentioned in Section 1.2.

2.6 Exact MISE calculations

The simplicity of the AMISE$\{\hat{f}(\cdot;h)\}$ formula given by (2.12) has many advantages. It allows a direct appreciation of the role of the bandwidth in the variance-bias trade-off that arises in nonparametric density estimation. In Sections 2.7 and 2.9 we will see that this approximation is useful for understanding the influence of the kernel and the true density on the performance of the kernel density estimator. However, it should be remembered that AMISE$\{\hat{f}(\cdot;h)\}$ is only a large sample approximation to MISE$\{\hat{f}(\cdot;h)\}$ given by (2.8). If one wants to analyse the exact finite sample performance of the kernel density estimator for a particular K and f then one could compute MISE$\{\hat{f}(\cdot;h)\}$ using (2.8). The biggest problem that arises with this is that (2.8) involves several integrals which, in general, would need to be computed numerically. However, there is a wide range of choices of K and f for which (2.8) can be computed *exactly*. This means that finite sample MISE$\{\hat{f}(\cdot;h)\}$ analyses can be greatly simplified. Recall the notation $\phi_\sigma(x) = (2\pi\sigma^2)^{-1/2}e^{-x^2/(2\sigma^2)}$. Then $\phi_\sigma(x-\mu)$ is the density of the $N(\mu,\sigma^2)$ distribution. A very useful formula for obtaining exact MISE expressions for a wide class of densities is

$$\int \phi_\sigma(x-\mu)\phi_{\sigma'}(x-\mu')\,dx = \phi_{(\sigma^2+\sigma'^2)^{1/2}}(\mu-\mu'). \qquad (2.15)$$

There are several ways to derive this result, but one straightforward way is to note that it follows directly from the algebraic identity

$$\phi_\sigma(x-\mu)\phi_{\sigma'}(x-\mu')$$
$$= \phi_{(\sigma^2+\sigma'^2)^{1/2}}(\mu-\mu')\phi_{\sigma\sigma'/(\sigma^2+\sigma'^2)^{1/2}}(x-\mu^*)$$

where $\mu^* = (\sigma'^2\mu + \sigma^2\mu')/(\sigma^2+\sigma'^2)$. Then, for appropriate choice of K and f, exact MISE expressions can be derived very simply from

$$\text{MISE}\{\hat{f}(\cdot;h)\} = (nh)^{-1}\int K^2(x)\,dx$$
$$+ (1-n^{-1})\int\left\{\int K_h(x-y)f(y)\,dy\right\}^2 dx \qquad (2.16)$$
$$- 2\int\left\{\int K_h(x-y)f(y)\,dy\right\}f(x)\,dx + \int f(x)^2\,dx$$

2.6 EXACT MISE CALCULATIONS

by using repeated application of (2.15). Note that we have replaced $(K_h * f)(x)$ by its definition in (2.8) to obtain (2.16).

EXAMPLE. Take K to be the $N(0,1)$ density and f to be the $N(0,\sigma^2)$ density. Then we have

$$K_h(x) = \phi_h(x) \quad \text{and} \quad f(x) = \phi_\sigma(x).$$

Using (2.15) we have

$$\int K^2(x)\,dx = \int \phi^2(x)\,dx = \phi_{2^{1/2}}(0) = (2\pi^{1/2})^{-1}.$$

The inner integral of the second and third terms of (2.16) is

$$\int K_h(x-y)f(y)\,dy = \int \phi_h(y-x)\phi_\sigma(y)\,dy = \phi_{(h^2+\sigma^2)^{1/2}}(x).$$

Note that this is the expected value of $\hat{f}(x;h)$. That is,

$$E\hat{f}(x;h) = \phi_{(h^2+\sigma^2)^{1/2}}(x).$$

The integral of the second term therefore becomes

$$\int \phi_{(h^2+\sigma^2)^{1/2}}(x)^2\,dx = \phi_{(2h^2+2\sigma^2)^{1/2}}(0)$$
$$= (2\pi^{1/2})^{-1}(\sigma^2 + h^2)^{-1/2}.$$

We can also simplify the other integrals in the same way and combine them to obtain

$$2\pi^{1/2}\text{MISE}\{\hat{f}(\cdot;h)\} = (nh)^{-1} + (1 - n^{-1})(\sigma^2 + h^2)^{-1/2}$$
$$+ \sigma^{-1} - 2^{3/2}(2\sigma^2 + h^2)^{-1/2}$$

(Fryer, 1976, Deheuvels, 1977a). ∎

It is easy to see that the calculations performed in the above example can be applied to any density that can be expressed as a finite linear combination of normal densities. Such densities can be written in the form

$$f(x) = \sum_{\ell=1}^{k} w_\ell \phi_{\sigma_\ell}(x - \mu_\ell) \tag{2.17}$$

where k is a positive integer, $w_1, \ldots, w_k$ is a set of positive numbers that sum to one, and for each ℓ, $-\infty < \mu_\ell < \infty$ and $\sigma_\ell^2 > 0$. This is called the family of *normal mixture densities*. We will leave it as an exercise (Exercise 2.10) to show that for f given by (2.17) and K the $N(0,1)$ density

$$\text{MISE}\{\hat{f}(\cdot;h)\} = (2\pi^{1/2}nh)^{-1} \\ + \mathbf{w}^T\{(1-n^{-1})\boldsymbol{\Omega}_2 - 2\boldsymbol{\Omega}_1 + \boldsymbol{\Omega}_0\}\mathbf{w} \tag{2.18}$$

(Marron and Wand, 1992) where $\mathbf{w} = (w_1 \cdots w_k)^T$ and $\boldsymbol{\Omega}_a$ is the $k \times k$ matrix having (ℓ, ℓ') entry equal to

$$\phi_{(ah^2+\sigma_\ell^2+\sigma_{\ell'}^2)^{1/2}}(\mu_\ell - \mu_{\ell'}).$$

The family of normal mixture densities is extremely rich (see Marron and Wand, 1992) and, in fact, any density can be approximated arbitrarily well by a member of this family. Therefore (2.18) allows finite sample analyses of kernel density estimators for a wide variety of situations. The flexibility is illustrated by Figure 2.4 where six different normal mixture densities are plotted.

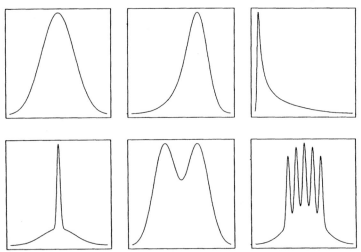

Figure 2.4. *Six normal mixture densities.*

Exact MISE expressions can be obtained for certain other kernels and densities (see e.g. Exercise 2.19).

Figure 2.5 shows $\text{MISE}\{\hat{f}(\cdot;h)\}$ (solid curve) versus $\log_{10} h$ for the normal mixture density f_1 when $n = 100$. For comparison purposes the plot of AMISE$\{\hat{f}(\cdot;h)\}$ (dashed curve) is also shown. Also included are plots of the integrated variance and its asymptotic approximation (the decreasing curves) and the integrated squared bias and its asymptotic approximation (the increasing curves). Vertical lines are drawn through the respective minimisers of MISE and AMISE. Their values are $h_{\text{MISE}} = 0.318$ and $h_{\text{AMISE}} = 0.246$ respectively. Notice that the integrated variance approximation is quite good for all values of h considered and is "uniform" in the sense that the vertical discrepancy tends to remain the same. On the other hand, the integrated squared bias approximation is very "non-uniform" and it worsens considerably as h becomes large. This is because the bias approximation is based on the assumption that $h \to 0$. This pattern seems to be typical for most densities, although the bias approximation can be considerably worse. Overall, for densities close to normality the bias approximation tends to be quite reasonable, while for densities with more features such as multiple modes this approximation becomes worse (see Marron and Wand, 1992).

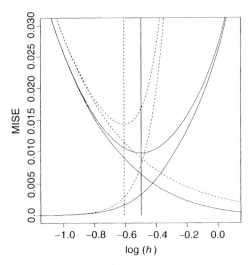

Figure 2.5. *Plots of* MISE$\{\hat{f}(\cdot;h)\}$ *(bowl-shaped solid curve) and* AMISE$\{\hat{f}(\cdot;h)\}$ *(bowl-shaped dashed curve) versus* $\log_{10} h$ *for the density* f_1 *and* $n = 100$. *Vertical lines are drawn through their respective minimisers. Also plotted are the integrated variance and its asymptotic approximation (decreasing solid and dashed curves respectively) and the integrated squared bias and its asymptotic approximation (increasing solid and dashed curves respectively).*

2.7 Canonical kernels and optimal kernel theory

We have already emphasised the crucial role of the bandwidth h and will defer its choice to Chapter 3. In this section, we investigate what effect the shape of the kernel function K has. Almost always K is taken to be a symmetric, unimodal density such as the normal density which we have used in the examples presented so far. This is often based on simplicity of interpretation, but there are also some compelling theoretical reasons for insisting that K be symmetric and unimodal, since there is a sense in which density estimators based on kernels that do not satisfy these requirements are inadmissible (Cline, 1988). Of course, there are many kernel functions that do satisfy these basic requirements, so it is worth asking whether certain kernel shapes are better than others. To answer this question we will first derive a useful rescaling of a kernel that allows a more appropriate comparison of kernel shapes.

Consider the formula for AMISE$\{\hat{f}(\cdot;h)\}$ given by (2.12). Optimisation of this expression with respect to K is not a clear-cut problem since the scaling of K is coupled with the bandwidth h. However, since the scaling of K is arbitrary we have the freedom to choose a rescaling of K of the form

$$K_\delta(\cdot) = K(\cdot/\delta)/\delta$$

that leads to a factorization of (2.12) into two terms that separate the dependence of h and K. This can be achieved by choosing δ so that

$$R(K_\delta) = \mu_2(K_\delta)^2. \qquad (2.19)$$

It is easily shown (Exercise 2.12) that (2.19) is satisfied by taking δ equal to

$$\delta_0 = \{R(K)/\mu_2(K)^2\}^{1/5}. \qquad (2.20)$$

Then

$$\text{AMISE}\{\hat{f}(\cdot;h)\} = C(K_{\delta_0})\{(nh)^{-1} + \tfrac{1}{4}h^4 R(f'')\} \qquad (2.21)$$

where

$$C(K) = \{R(K)^4 \mu_2(K)^2\}^{1/5}.$$

Notice that $C(K)$ is invariant to rescalings of K. That is, $C(K_{\delta_1}) = C(K_{\delta_2})$ for any $\delta_1, \delta_2 > 0$ (Exercise 2.14).

We call $K^c = K_{\delta_0}$ the *canonical kernel* for the class $\{K_\delta : \delta > 0\}$ of rescalings of K since it is the unique member of that class that

2.7 CANONICAL KERNELS AND OPTIMAL KERNEL THEORY

permits the "decoupling" of K and h apparent in (2.21) (Marron and Nolan, 1989).

EXAMPLE. Let $K = \phi$, the standard normal kernel. Then using (2.15) we obtain $\delta_0 = (4\pi)^{-1/10}$ so the canonical kernel for the class $\{\phi_\delta : \delta > 0\}$ is

$$\phi^c(x) = \phi_{(4\pi)^{-1/10}}(x)$$

and $C(\phi) = (4\pi)^{-2/5}$. If $\hat{f}^c(x; h) = n^{-1} \sum_{i=1}^{n} \phi_h^c(x - X_i)$ is the kernel density estimator based on ϕ^c then

$$\text{AMISE}\{\hat{f}^c(\cdot; h)\} = (4\pi)^{-2/5}\{(nh)^{-1} + \tfrac{1}{4}h^4 R(f'')\}.$$

∎

Canonical kernels are useful for pictorial comparison of density estimates based on different shaped kernels since they are defined in such a way that a particular single choice of bandwidth gives roughly the same amount of smoothing (see Marron and Nolan, 1989). This is illustrated in Figure 2.6.

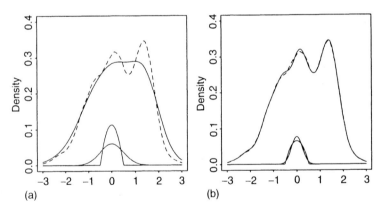

Figure 2.6. (a) Kernel density estimates based on equal bandwidths but different kernels (solid curve is for standard normal kernel, dashed curve is for K^*). (b) Kernel density estimates based on equal bandwidths and different canonical kernels. The small curves at the base of the graph represent the kernel mass for each estimate.

Figure 2.6 (a) shows kernel density estimates based on the standard normal kernel (solid curve) and kernel K^* as defined by (2.22) below (dashed curve). The same bandwidth is used for each estimate, but very different estimates result. In Figure 2.6 (b) estimates based on the canonical versions of these kernels and equal bandwidths are plotted. In this case the estimates are almost identical.

The problem of determining the optimal kernel shape can be handled more easily using canonical kernels since, by (2.21), we only need to choose K to minimise $C(K_{\delta_0})$. However, because of the scale invariance of $C(K)$, the optimal K is the one that minimises $C(K)$ subject to

$$\int K(x)\,dx = 1, \quad \int xK(x)\,dx = 0, \quad \int x^2 K(x)\,dx = a^2 < \infty$$

and $K(x) \geq 0$ for all x.

The solution can be shown to be

$$K^a(x) = \tfrac{3}{4}\{1 - x^2/(5a^2)\}/(5^{1/2}a)1_{\{|x|<5^{1/2}a\}}$$

(Hodges and Lehmann, 1956). Of course, a is an arbitrary scale parameter and the simplest version of K^a corresponds to $a^2 = \tfrac{1}{5}$ which yields

$$K^*(x) = \tfrac{3}{4}(1 - x^2)1_{\{|x|<1\}}. \tag{2.22}$$

This kernel is often called the *Epanechnikov kernel* since its optimality properties in the density estimation setting were first described by Epanechnikov (1969). Its graph is shown in Figure 2.7.

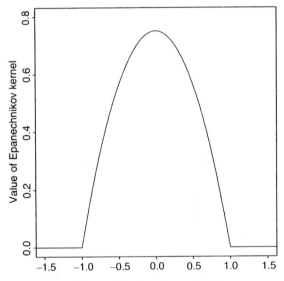

Figure 2.7. *The Epanechnikov kernel* K^*.

2.7 CANONICAL KERNELS AND OPTIMAL KERNEL THEORY

Table 2.1 shows values of $\{C(K^*)/C(K)\}^{5/4}$ for various popular kernels K. This ratio is sometimes called the *efficiency* of K relative to K^* since it represents the ratio of sample sizes necessary to obtain the same minimum AMISE (for a given f) when using K^* as when using K. For example, if K has an efficiency of 0.95 this indicates that the density estimate optimal kernel K^* can achieve the same minimum AMISE using 95% of the data as that using K. Four of the kernels are particular cases of the family

$$K(x;p) = \{2^{2p+1}B(p+1,p+1)\}^{-1}(1-x^2)^p 1_{\{|x|<1\}},$$

where $B(\cdot,\cdot)$ is the beta function. These kernels are symmetric beta densities on the interval $[-1,1]$. Note that $p = 0$ gives the uniform density, $p = 1$ is the Epanechnikov kernel, and the cases $p = 2$ and $p = 3$ are often called *biweight* and *triweight* kernels, respectively. The fourth kernel in the table is the standard normal density ϕ, although it is related to the beta family as their limiting case as $p \to \infty$. The triangular kernel is given by $K(x) = (1-|x|)1_{\{|x|<1\}}$.

Table 2.1. *Efficiencies of several kernels compared to the optimal kernel.*

Kernel	$\{C(K^*)/C(K)\}^{5/4}$
Epanechnikov	1.000
Biweight	0.994
Triweight	0.987
Normal	0.951
Triangular	0.986
Uniform	0.930

The main message from Table 2.1 is that the "suboptimal" kernels are not suboptimal by very much, and hence one loses very little in performance terms. Indeed, these results suggest that most unimodal densities perform about the same as each other when used as a kernel. It follows that choice between kernels can be made on other grounds such as computational efficiency. Uniform kernels are not very popular in practice since the corresponding density estimate is piecewise constant, and even the Epanechnikov kernel gives an estimate having a discontinuous first derivative which can sometimes be unattractive because of its "kinks".

2.8 Higher-order kernels

In Section 2.5 we showed that the best obtainable rate of convergence of the kernel estimator considered there is of order $n^{-4/5}$. In this section we demonstrate that it is possible to obtain better rates of convergence by relaxing the restriction that the kernel be a density function.

Recall the result for the asymptotic bias given by (2.10),

$$E\hat{f}(x;h) - f(x) = \tfrac{1}{2}h^2\mu_2(K)f''(x) + o(h^2).$$

When K is constrained to be a probability density function then it is necessarily true that $\mu_2(K) > 0$. However, without this restriction, it is possible to construct K so that $\mu_2(K) = 0$ which will have the effect of reducing the bias to be of order h^4, provided assumption (i) in Section 2.6 is strengthened to f having a continuous square integrable fourth derivative. It is easy to check that the MSE and MISE will have optimal rate of convergence of order $n^{-8/9}$ if such a kernel is used.

We can extend this idea further by considering kernels which have several vanishing moments. Define

$$\mu_j(K) = \int x^j K(x)\,dx$$

to be the jth moment of the kernel K. Then we will say that K is a *kth-order* kernel if

$$\mu_0(K) = 1, \; \mu_j(K) = 0 \text{ for } j = 1,\ldots,k-1, \text{ and } \mu_k(K) \neq 0.$$

We will still require that K be symmetric, which implies that k is even.

There are several rules for constructing higher-order kernels (e.g. Jones and Foster, 1993). Let $K_{[\ell]}$ denote an ℓth-order kernel. Then formulae such as

$$K_{[k+2]}(x) = \tfrac{3}{2}K_{[k]}(x) + \tfrac{1}{2}xK'_{[k]}(x) \qquad (2.23)$$

can be used to generate higher-order kernels (here $K_{[k]}(x)$ is assumed to be differentiable). For example, application of (2.23) to the second-order kernel $K_{[2]}(x) = \phi(x)$ leads to the fourth-order kernel

$$K_{[4]}(x) = \tfrac{1}{2}(3 - x^2)\phi(x). \qquad (2.24)$$

2.8 HIGHER-ORDER KERNELS

Figure 2.8 plots these two kernels together, with each scaled to have the value of 1 at the origin. Notice the negative lobes of $K_{[4]}$ which entail that the resulting density estimate is not a density itself.

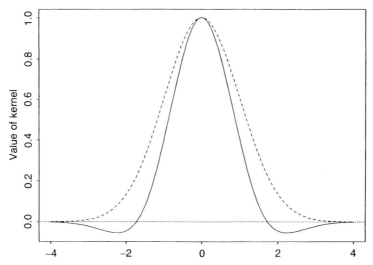

Figure 2.8. *Plot of $K_{[2]}(x) = \phi(x)$ (dashed curve) and $K_{[4]}(x)$ given by (2.24) (solid curve).*

Suppose now that $\hat{f}(\cdot; h)$ is based on a kth-order kernel $K_{[k]}$ and the density f has a kth continuous square integrable derivative. Then

$$E\hat{f}(x; h) = \int K_{[k]}(z) \sum_{\ell=0}^{k} (-hz)^{\ell}(\ell!)^{-1} f^{(\ell)}(x)\, dz + o(h^k)$$

$$= f(x) + (-1)^k \{\mu_k(K_{[k]})/k!\} h^k f^{(k)}(x) + o(h^k).$$

This leads to the AMISE expression

$$\mathrm{AMISE}\{\hat{f}(\cdot; h)\} = (nh)^{-1} R(K_{[k]}) + h^{2k} \{\mu_k(K_{[k]})/k!\}^2 R(f^{(k)}).$$

The AMISE-optimal bandwidth is then

$$h_{\mathrm{AMISE}} = \left[\frac{(k!)^2 R(K_{[k]})}{2k\mu_k(K_{[k]})^2 R(f^{(k)}) n} \right]^{1/(2k+1)}$$

and

$$\inf_{h>0} \mathrm{MISE}\{\hat{f}(\cdot; h)\} \qquad (2.25)$$

$$\sim \frac{2k+1}{2k} \left\{ 2k(k!)^{-2} \mu_k(K_{[k]})^2 R(K_{[k]})^{2k} R(f^{(k)}) n^{-2k} \right\}^{1/(2k+1)}.$$

Since $2k/(2k+1)$ approaches 1 as k becomes larger (2.25) implies that, for sufficiently smooth densities, the convergence rate can be made arbitrarily close to the parametric n^{-1} convergence rate. The fact that higher-order kernels can achieve improved rates of convergence means that they will eventually dominate second-order kernel estimators for large n. However, this does not mean that a higher-order kernel will necessarily improve the error for sample sizes usually encountered in practice and, in many cases, unless the sample size is very large there will actually be an increase in the error due to using higher-order kernels. For example, if one decides to use the fourth-order kernel (2.24) as opposed to the standard normal kernel and the true density is the normal mixture f_1, then the effect on the kernel-dependent constants, particularly the value of $R(K_{[4]})$, is such that one will require a sample size of several thousand before there is a significant reduction in the MISE. Similar results hold for several other density shapes (see Marron and Wand, 1992, Table 2). Higher-order kernels necessarily take on negative values so there is also a price to be paid in terms of interpretability and plausibility. It is more difficult to understand the averaging process when there are negative contributions and the resulting density estimate will not be a density itself. Therefore, one needs to decide whether the practical gains of higher-order kernels are significant enough to sacrifice the simplicity of second-order kernels.

It is also possible to obtain explicit MISE expressions when f is a normal mixture density for a certain class of higher-order kernels (Marron and Wand, 1992). These kernels are higher-order extensions of the second-order normal kernel and are given by

$$G_{[k]}(x) = \sum_{\ell=0}^{k/2-1} \frac{(-1)^\ell}{2^\ell \ell!} \phi^{(2\ell)}(x), \quad \ell = 0, 2, 4, \ldots \qquad (2.26)$$

(Deheuvels, 1977a; Wand and Schucany, 1990). One can show (Exercise 2.16) that $G_{[k]}$ is, in fact, a kth-order kernel. In Figure 2.9, we plot $\log_{10}\{\inf_{h>0} \text{MISE}(h)\}$ as a function of $\log_{10} n$ for the density f_1 and the Gaussian based kernels $G_{[k]}$ for $k = 2, 4, 6$ and 8. By plotting on a log-log scale, each curve tends to a straight line with slope $-2k/(2k+1)$ as n gets large. But one can also see from this (a) the lack of improvement using higher-order kernels for small n and (b) the relative merits of each increase in order.

2.8 HIGHER-ORDER KERNELS

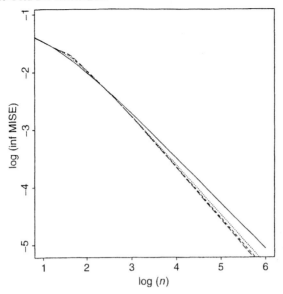

Figure 2.9. $\log_{10}\{\inf_{h>0} \text{MISE}(h)\}$ plotted against $\log_{10} n$ for density f_1 and Gaussian based kernels $G_{[k]}$ for $k = 2$ (solid curve), 4 (dotted curve), 6 (short-dashed curve) and 8 (long-dashed curve).

A natural extension of higher-order kernels is to those which have "infinite" order, that is, $\mu_j(K) = 0$ for all $j = 1, 2, \ldots$. The simplest such kernel is the "sinc" kernel

$$K(x) = \sin x / (\pi x) \tag{2.27}$$

(Davis, 1975). The MISE rates of convergence for the sinc kernel estimator depend on the tail behaviour of the characteristic function of f. For example, if f is the normal density it can be shown that for K given by (2.27)

$$\inf_{h>0} \text{MISE}\{\hat{f}(\cdot; h)\} = O\{(\log n)^{1/2} n^{-1}\}$$

which is faster than any rate of order $n^{-\alpha}$, $0 < \alpha < 1$. The MISE is not $O(n^{-1})$ because $R(K)$ is infinite for the sinc kernel. The sinc kernel estimator suffers from the same drawbacks as other higher-order kernel estimators and the good asymptotic performance is not guaranteed to carry over to finite sample sizes in practice.

2.9 Measuring how difficult a density is to estimate

One inherent difficulty with the kernel density estimator is that, while it provides good estimates for many density shapes, it can be inadequate for other shapes. A major source of this inadequacy is that the kernel estimator has just a single smoothing parameter which is used over the entire real line. This difficulty is illustrated in Figure 2.10 which shows kernel estimates of the lognormal density $f(x) = \phi(\ln x)/x$, when $n = 1000$ and the standard normal kernel is used. The solid curve in Figure 2.10 (a) is based on the bandwidth $h = 0.05$ which is chosen for good estimation of the mode. Unfortunately, such a small bandwidth leads to a very undersmoothed estimate in the tail with many spurious wiggles being introduced. If a larger bandwidth $h = 0.45$ is used, as shown in Figure 2.10 (b), the estimate in the tail region becomes better, yet the mode is now oversmoothed, as the dashed curve shows. The dashed curve of Figure 2.10 (c) shows the kernel estimate based on the intermediate bandwidth $h = 0.15$ with the hope of achieving a compromise between correct smoothing of the mode and of the tail; however, we see that such a choice is still unsatisfactory. It appears that adequate estimation of the lognormal density via (2.1) is quite difficult to achieve, unless n is extremely large.

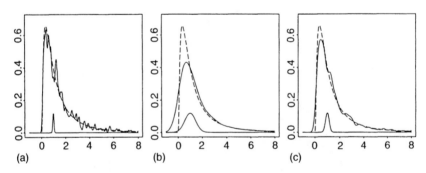

Figure 2.10. *Kernel estimates (solid curves) of the lognormal density (dashed curve) with $n = 1000$ and the standard normal kernel. Bandwidths are (a) $h = 0.05$, (b) $h = 0.45$ and (c) $h = 0.15$. The small curves at the base of the graph represent the kernel mass for each estimate.*

Our aim here is to quantify how well a particular density can be estimated using the kernel density estimator. In the next section we will look at ways of improving the performance of this estimate

2.9 MEASURING HOW DIFFICULT A DENSITY IS TO ESTIMATE

to handle more "difficult" density shapes. Recall that, for K a symmetric probability density function,

$$\inf_{h>0} \text{MISE}\{\hat{f}(\cdot;h)\} \sim \tfrac{5}{4}C(K)R(f'')^{1/5}n^{-4/5}, \qquad (2.28)$$

so that the dependence on f in the leading term is through the functional $R(f'') = \int f''(x)^2\,dx$. The magnitude of this quantity gives us an indication of how well f can be estimated even when h in $\hat{f}(\cdot;h)$ is chosen optimally. We already noted that $R(f'')$ was a measure of total curvature in Section 2.6. From (2.28) we see that this means that if f has "sharp" features such as high skewness or several modes then $|f''(x)|$ will take on relatively large values and result in a larger value of $R(f'')$. For densities without such features, the density estimation problem is easier because $R(f'')$ is lower.

Our first goal is to determine which density shape is the easiest for the kernel estimator to estimate in terms of having the smallest $R(f'')$ value. A first answer might be to take $f''(x) = 0$ everywhere so that $R(f'') = 0$. This would imply that f be linear; however, there is no density which is linear across the whole real line. The uniform density on a known finite interval is an obvious candidate for good performance, but with such a density there are problems associated with estimation near its endpoints. We confine further attention here to densities with a continuous square integrable second derivative over the whole real line. Before we optimise $R(f'')$ it should be noted that this functional is not scale invariant so we must fix scale in some way before carrying out the minimisation, otherwise any value for $R(f'')$ can be attained just by varying scale. To appreciate this, suppose that X has density f_X and $Y = X/a$ for some $a > 0$. Then, noting that the density of Y is $f_Y(x) = af_X(ax)$ we obtain, after some simple calculus,

$$R(f_Y'') = a^5 R(f_X'').$$

This shows that, while $R(f'')$ is not scale invariant, it is a simple matter to check that the difficulty measure

$$D(f) = \{\sigma(f)^5 R(f'')\}^{1/4},$$

where $\sigma(f)$ is the population standard deviation of f, is scale invariant. The inclusion of the $\tfrac{1}{4}$ power allows a simple equivalent sample size interpretation as was done for the comparison of kernel shapes in Section 2.8. Therefore $D(f)$ is a reasonable measure of the "degree of difficulty" of kernel estimation of f. The standard deviation is one of several scale measures that could be used in the definition of $D(f)$ (see Terrell, 1990, for others) and is chosen largely for simplicity. However, $D(f)$ is not appropriate for measuring the difficulty of estimation of all types of densities such as those having two widely separated modes (see Exercise 3.2). It can be shown (Terrell, 1990) that $D(f)$ is minimal when

$$f^*(x) = \tfrac{35}{32}(1-x^2)^3 1_{\{|x|<1\}},$$

the beta (4,4) or triweight density, and that the minimum value is 35/243. Note that we really have an equivalence class of densities achieving this minimum value of $D(f)$, since any shift or rescaling of f^* will also optimise $D(f)$. Thus, in terms of $D(f)$, the density f^* shown in Figure 2.11 is the easiest density to estimate.

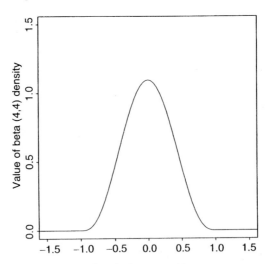

Figure 2.11. *The beta (4,4) density f^*.*

By analogy with what was done for kernel functions in Section 2.7, we can use $D(f)$ to compare the performance of $\hat{f}(\cdot;h)$ on density f with what it achieves for density f^*. The efficiency of the kernel estimator for estimating density f compared to estimating the easiest density f^* can be defined by $D(f^*)/D(f)$, values of which are given for several shapes of densities in Table 2.2. These densities are plotted in Figure 2.12. The gamma(3) density is defined in Table B.1 of Appendix B. The lognormal density is as defined in the first paragraph of this section.

2.9 MEASURING HOW DIFFICULT A DENSITY IS TO ESTIMATE

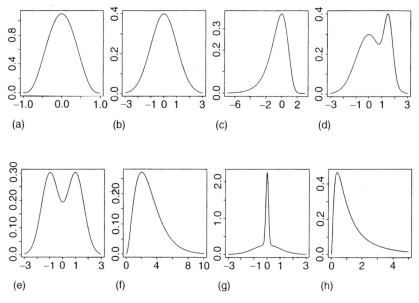

Figure 2.12. *Graphs of each density in Table 2.2*

Table 2.2. *Values of $D(f^*)/D(f)$ for several densities*

Density	$D(f^*)/D(f)$
(a) Beta(4,4)	1
(b) Normal	0.908
(c) Extreme value	0.688
(d) $\frac{3}{4}N(0,1) + \frac{1}{4}N(\frac{3}{2},(\frac{1}{3})^2)$	0.568
(e) $\frac{1}{2}N(-1,\frac{4}{9}) + \frac{1}{2}N(1,\frac{4}{9})$	0.536
(f) Gamma(3)	0.327
(g) $\frac{2}{3}N(0,1) + \frac{1}{3}N(0,\frac{1}{100})$	0.114
(h) Lognormal	0.053

Notice that densities close to normality appear to be easiest for the kernel estimator to estimate. The degree of estimation difficulty increases with skewness, kurtosis and multimodality. In terms of the $D(f)$ criterion the last table value shows that more than fifteen times as many data are required to estimate the lognormal density with the same accuracy as an estimate of the normal density, which is in accordance with the discussion given at the beginning of this section.

2.10 Modifications of the kernel density estimator

Section 2.9 suggests that while the basic kernel density estimator works well in some situations, there is scope for improvement in others. In this section we will look at some adaptations of the kernel density estimator which attempt to overcome its limitations. At the time of writing none of these modifications had become a widely accepted method. The main reason for this is that, for each of them, there are still many unsettled issues concerning performance and practical implementation. Therefore, we will restrict our presentation to the fundamental principles and properties of each estimator. For more recent developments in the literature we refer the reader to the bibliographical notes at the end of this chapter.

2.10.1 *Local kernel density estimators*

The kernel density estimator $\hat{f}(\cdot;h)$ defined in (2.1) depends on a single smoothing parameter h. Given that the optimal amount of smoothing varies across the real line, an obvious extension of (2.1) is to that having a different bandwidth $h(x)$, say, for each point x at which $f(x)$ is estimated.

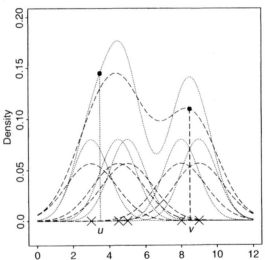

Figure 2.13. *How $\hat{f}_L$ is made up: at u the dashed kernels, with the larger bandwidth $h(u)$, are averaged to form $\hat{f}_L(u;h(u))$, while $\hat{f}_L(v;h(v))$ is the average of the narrower dotted kernels associated with the smaller $h(v)$.*

2.10 MODIFICATIONS OF THE KERNEL DENSITY ESTIMATOR

This leads to the *local* kernel density estimator given by

$$\hat{f}_L(x; h(x)) = \{nh(x)\}^{-1} \sum_{i=1}^{n} K\{(x - X_i)/h(x)\}. \quad (2.29)$$

Notice that $\hat{f}_L$ works by employing a different basic kernel estimator at each point. This is illustrated in Figure 2.13 in which the values of $\hat{f}_L$ at distinct points u and v are seen to arise from the dashed and dotted kernels, respectively. A consequence of allowing the bandwidth to depend on x is that $\hat{f}_L(\cdot; h(\cdot))$ is usually not a density function itself.

The analogue of (2.13) for asymptotic MSE at x is

$$h_{\text{AMSE}}(x) = \left[\frac{R(K)f(x)}{\mu_2(K)^2 f''(x)^2 n}\right]^{1/5},$$

provided $f''(x) \neq 0$. (When $f''(x) = 0$, further terms in the asymptotic expansion of MSE must be taken into account; see Schucany, 1989). If this choice of $h_{\text{AMSE}}(x)$ is made for each x, then the corresponding value of $\text{AMISE}\{\hat{f}_L(\cdot; h(\cdot))\}$ can be shown to be

$$\tfrac{5}{4}\{\mu_2(K)^2 R(K)^4\}^{1/5} R((f^2 f'')^{1/5}) n^{-4/5}.$$

Notice that the rate of convergence of $\hat{f}_L$ is $n^{-4/5}$ which means that there is no improvement in these terms. However, it can be shown (Exercise 2.22) that

$$R((f^2 f'')^{1/5}) \leq R(f'')^{1/5} \quad (2.30)$$

for all f so there will always be some improvement if $h(x)$ is chosen optimally. In terms of sample size efficiency this improvement depends on the magnitude of the ratio

$$\{R((f^2 f'')^{1/5})/R(f'')^{1/5}\}^{5/4}.$$

In practice, (2.29) requires specification of the whole function $h(x)$ before the main smoothing stage can be done. Preliminary estimation of a function like this is usually called *pilot estimation*. We will just mention one popular method which fits into this framework. The *nearest neighbour density estimator* (Loftsgaarden and Quesenberry, 1965) uses distances from x to the data point that is the kth nearest to x (for some suitable k) in a pilot estimation step that is essentially equivalent to $h(x) \propto 1/f(x)$. However, it is not very difficult to see that there are many situations where $1/f(x)$ is an unsatisfactory surrogate for the optimal $\{f(x)/f''(x)^2\}^{1/5}$.

2.10.2 Variable kernel density estimators

A quite different idea from local kernel density estimation is that of *variable* kernel density estimation. In the variable kernel density estimator, the single h is replaced by n values $\alpha(X_i)$, $i=1,\ldots,n$, rather than by $h(x)$. This estimator is therefore of the form

$$\hat{f}_V(x;\alpha) = n^{-1}\sum_{i=1}^{n}\{\alpha(X_i)\}^{-1}K\{(x-X_i)/\alpha(X_i)\}. \qquad (2.31)$$

The idea here is that the kernel centred on X_i has associated with it its own scale parameter $\alpha(X_i)$, thus allowing different degrees of smoothing depending on where X_i is in relation to other data points. This is illustrated in Figure 2.14. In particular, the aim is to smooth out the mass associated with data values that are in sparse regions much more than those situated in the main body of the data. Also, if K is a probability density then so is $\hat{f}_V(\cdot;\alpha)$.

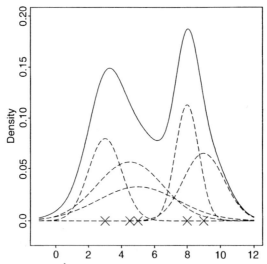

Figure 2.14. *How $\hat{f}_V(\cdot;\alpha)$ is made up, as the average of kernels with different scalings.*

The above intuition suggests that each $\alpha(X_i)$ should depend on the true density in roughly an inverse way. In fact, theoretical work by Abramson (1982) shows that taking $\alpha(X_i) = hf^{-1/2}(X_i)$ is a particularly good choice because then one can achieve a bias of order h^4 rather than h^2, sufficient differentiability of f permitting. (This actually needs, at least in theory, an additional "clipping" procedure which stops values of $\alpha(X_i)$ becoming too large in areas

2.10 MODIFICATIONS OF THE KERNEL DENSITY ESTIMATOR

where f is small, but such technicalities are beyond the scope of this book). Pilot estimation to obtain $\alpha(X_i)$, $i = 1, \ldots, n$, is necessary for a practical implementation of $\hat{f}_V(\cdot; \alpha)$.

We will not derive the MSE for $\hat{f}_V$ in this book, but we will take one formula from the literature to illustrate $\hat{f}_V(\cdot; \alpha)$'s effects. Suppose that $\alpha(X_i) = hf^{-1/2}(X_i)$. If a kernel estimate with bandwidth an order of magnitude smaller than h is employed as pilot then

$$\text{AMSE}\{\hat{f}_V(x; \alpha)\} = (nh)^{-1}S(K)f^{3/2}(x)$$
$$+ \tfrac{1}{576}h^8\mu_4(K)^2(1/f)''''(x)^2$$

(Hall and Marron, 1988). Here, $S(K) = \{6R(K) + R(xK')\}/4$. With respect to rates of convergence, this MSE is of the same form as that associated with a fourth-order kernel in Section 2.8. The optimal h is of order $n^{-1/9}$ and the optimal MSE is improved from order $n^{-4/5}$ to $n^{-8/9}$. This is achieved at the expense of pilot estimation but without any negativity, when compared with using a fourth-order kernel.

2.10.3 Transformation kernel density estimators

If the random sample $X_1, \ldots, X_n$ has a density f that is difficult to estimate then another possibility is to apply a transformation to the data to obtain a new sample $Y_1, \ldots, Y_n$ having a density g that can be more easily estimated using the basic kernel density estimator. One would then "backtransform" the estimate of g to obtain the estimate of f. The resulting estimator is called the *transformation kernel density estimator*.

Suppose that our transformation is given by $Y_i = t(X_i)$ where t is an increasing differentiable function defined on the support of f. Then a standard result from statistical distribution theory is that

$$f(x) = g(t(x))t'(x)$$

The transformation kernel density estimator of f is obtained by replacing g by $\hat{g}(\cdot; h)$, the ordinary kernel density estimator of g based on $Y_1, \ldots, Y_n$ which leads to the explicit formula

$$\hat{f}_T(x; h, t) = n^{-1}\sum_{i=1}^{n} K_h\{t(x) - t(X_i)\}t'(x). \quad (2.32)$$

The transformation kernel density estimator is neither a local nor a variable kernel density estimator. However, application of the mean value theorem to (2.32) gives

$$\hat{f}_T(x; h, t) = n^{-1} \sum_{i=1}^{n} \{t'(x)/h\} K_h \{t'(\xi_i)(x - X_i)/h\},$$

where ξ_i lies between x and X_i which, in view of (2.29) and (2.31), shows that $\hat{f}_T(\cdot; h, t)$ has something in common with $\hat{f}_L(\cdot; h(\cdot))$ and $\hat{f}_V(\cdot; \alpha)$.

A simple illustrative example of this idea comes from the problem of estimating the lognormal density. As we saw in Section 2.9, this density is very difficult to estimate by direct kernel methods. However, if we apply the transformation $Y_i = \ln X_i$ to the data then the Y_i's are a sample from the $N(0, 1)$ distribution. As indicated by Table 2.2, the normal density is much easier to estimate than the lognormal. Figure 2.15 (a) shows an ordinary kernel estimate of the data of Section 2.9 after they have undergone the natural log transformation.

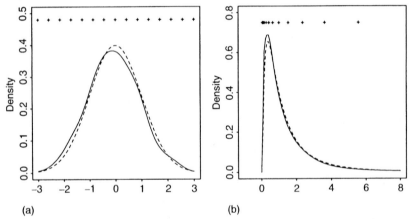

Figure 2.15. *Transformation kernel density estimate of simulated lognormal data ($n = 1000$). In (a) the solid curve is the ordinary kernel density estimate on the transformed scale, while the solid curve in (b) is its backtransformation using $t^{-1}(x) = e^x$. The dashed curves are the appropriate true functions. The plus signs show the effect of the transformation as described in the text.*

In Figure 2.15 (b) the estimate of the lognormal density is shown. This estimate is given by (2.32) with $t(x) = \ln x$, but can also

2.10 MODIFICATIONS OF THE KERNEL DENSITY ESTIMATOR

be thought of as the "backtransformation" of the kernel density estimate in Figure 2.15 (a) according to the inverse transformation $t^{-1}(x) = e^x$. The effect of this backtransformation is indicated by the plus signs at the top of each plot. These are equally spaced in Figure 2.15 (a). In Figure 2.15 (b) they have undergone the backtransformation.

The best choice of the transformation t depends quite heavily on the shape of f. If f is a skewed unimodal density then it can be argued (Wand, Marron and Ruppert, 1991) that t should be a convex function on the support of f since such a t will, in a certain sense, reduce the skewness of f. If f is close to being symmetric, but has a high amount of kurtosis, then t should be concave to the left and convex to the right of the centre of symmetry of f (Ruppert and Wand, 1992).

One approach to the selection of t is by restricting t to an appropriate parametric family of transformations. For heavily skewed data one possible two-parameter family of convex transformations is the *shifted power family* given by

$$t(x; \lambda_1, \lambda_2) = \begin{cases} (x + \lambda_1)^{\lambda_2} \operatorname{sign}(\lambda_2), & \lambda_2 \neq 0 \\ \ln(x + \lambda_1), & \lambda_2 = 0, \end{cases}$$

where $\lambda_1 > -\min(X)$ and $\min(X)$ denotes the lower endpoint of the support of f. This is an extension of the Box-Cox family of transformations (see Wand et al., 1991). The parameters could be estimated in such a way that g is as easy to estimate as possible using the ordinary kernel density estimator. One possible strategy is to choose the parameters to minimise an estimate of $D(g)$ (Wand et al., 1991). The estimate of the British household incomes data in Figure 2.1 (b) is a transformation kernel density estimate based on the shifted power family with $\lambda_1 = 0.0$ and $\lambda_2 = 0.26$.

An alternative approach is to estimate t nonparametrically. If F and G are the distribution functions of the densities f and g then a well known result is that $Y = G^{-1}(F(X))$ has density g. One could then choose G to correspond to an "easy to estimate" density and take $t = G^{-1}(\hat{F}(\cdot; h))$ where $\hat{F}(\cdot; h)$ is a kernel estimate of F (see Ruppert and Cline, 1994).

2.11 Density estimation at boundaries

Throughout most of this chapter we have assumed that the density f satisfies certain smoothness criteria. For example, the asymptotic MISE analysis presented in Section 2.5 requires that f have a continuous second derivative over the entire real line. There are, however, many common densities that do not satisfy this condition. A simple example is the exponential density $f(x) = e^{-x}$, $x > 0$. Figure 2.16 shows a kernel estimate of this density (solid curve) based on a sample size of $n = 1000$. The true density is shown for comparison (dashed curve). It is clear that the kernel estimation of f near its discontinuity is difficult. For x positive and close to zero the kernel estimator is trying to estimate relatively high density values, while for x negative and close to zero the density estimate is aiming to estimate zero. It seems clear that the continuous operation of kernel smoothing will not perform well at this discontinuity in the function being estimated.

To quantify the density estimation problem mathematically near a boundary, suppose that f is a density such that $f(x) = 0$ for $x < 0$ and $f(x) > 0$ for $x \geq 0$. We will suppose that f'' is continuous away from $x = 0$. Also, let K be a kernel with support confined to $[-1, 1]$.

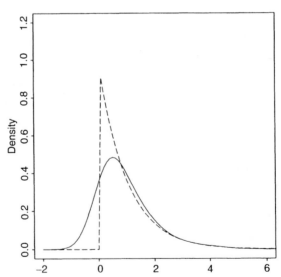

Figure 2.16. *Kernel estimate of exponential density based on a sample of $n = 1000$. The solid curve is the density estimate. The dashed curve is the true density.*

2.11 DENSITY ESTIMATION AT BOUNDARIES

For $x > 0$,

$$E\hat{f}(x; h) = \int_{-1}^{x/h} K(z) f(x - hz) \, dz. \tag{2.33}$$

If $x \geq h$ then the upper limit on the integral in (2.33) can be replaced by 1 and the analysis is the same as usual. For any fixed $x > 0$ and h converging to zero we will always have $x > h$ for sufficiently large n. However, important insight into the boundary behaviour of the kernel estimator can be obtained by studying its asymptotic properties at a sequence of points that converge to the boundary in such a way that each member of the sequence is always within one bandwidth of the boundary. The simplest way to do this is to take $x = x_n = \alpha h$ where $0 \leq \alpha < 1$. From (2.33) we obtain for such x

$$\begin{aligned} E\hat{f}(x; h) &= \int_{-1}^{\alpha} K(z) f(x - hz) \, dz \\ &= f(x)\nu_{0,\alpha}(K) - hf'(x)\nu_{1,\alpha}(K) \\ &\quad + \tfrac{1}{2} h^2 f''(x)\nu_{2,\alpha}(K) + O(h^3) \end{aligned}$$

where $\nu_{\ell,\alpha}(K) = \int_{-1}^{\alpha} z^\ell K(z) \, dz$. Since $\nu_{0,\alpha}(K) \neq 1$ in general we no longer have consistency at such points. At the boundary we obtain

$$E\hat{f}(0; h) = \tfrac{1}{2} f(0) + O(h).$$

This result is consistent with the intuitive idea of the kernel estimator having to find a compromise between estimating the two distinct values of f on either side of the discontinuity.

Since the location of the boundary is usually known, $\hat{f}(x; h)$ can be adapted to achieve better performance in its vicinity. An obvious first idea is to normalise $\hat{f}(x; h)$ by dividing by $\nu_{0,\alpha}(K)$ at each x. This achieves consistency near the boundary, but still results in a large $O(h)$ bias there. A variety of further modifications is possible to achieve $O(h^2)$ bias everywhere. One can think of these boundary modifications in terms of special "boundary kernels" which are different for each $x = \alpha h$ (Gasser and Müller, 1979). One simple family of boundary kernels is the following linear multiple of the kernel K:

$$K^L(u; \alpha) = \frac{\nu_{2,\alpha}(K) - \nu_{1,\alpha}(K) u}{\nu_{0,\alpha}(K)\nu_{2,\alpha}(K) - \nu_{1,\alpha}(K)^2} K(u) 1_{\{-1 < u < \alpha\}}$$

(e.g. Gasser and Müller, 1979). Figure 2.17 plots $K^L(u; \alpha)$ for several values of α, and K equal to the biweight kernel.

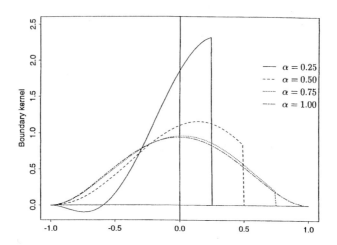

Figure 2.17. $K^L(u;\alpha)$ based on the biweight kernel for $\alpha = \frac{1}{4}$ (solid curve), $\frac{1}{2}$ (dashed curve), $\frac{3}{4}$ (dotted curve) and 1 (dot-dashed curve).

Figure 2.18 shows a kernel density estimate of the exponential density with $K^L(\cdot;\alpha)$ used near the boundary, with K equal to the biweight kernel. The sample size is $n = 1000$ and the bandwidth is $h = 0.5$.

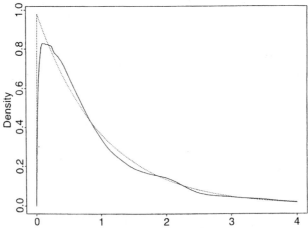

Figure 2.18. Kernel density estimate based on a sample of size 1000 from the exponential density with $K^L(\cdot;\alpha)$ used near the left boundary at $x = 0$. The kernel K is the biweight kernel with bandwidth $h = 0.5$.

Observe the improved performance near the boundary compared to the estimate in Figure 2.16.

2.12 Density derivative estimation

There are a number of reasons for wishing to estimate derivatives of f. First and second derivatives may be of intrinsic interest as measures of slope and curvature; other important functions, such as the score function $-f'/f$, depend on density derivatives; automatic bandwidth selection methods require estimation of quantities involving density derivatives (see Chapter 3); and estimates of modes and inflexion points require derivative estimates.

A natural estimator of the rth derivative $f^{(r)}(x)$ is

$$\hat{f}^{(r)}(x;h) = n^{-1}h^{-r-1}\sum_{i=1}^{n} K^{(r)}\{(x-X_i)/h\},$$

sufficient differentiability of K permitting. Mean squared error properties of $\hat{f}^{(r)}(x;h)$ can be derived straightforwardly (Exercise 2.25) to obtain

$$\begin{aligned}\text{MSE}\{\hat{f}^{(r)}(x;h)\} &= (nh^{2r+1})^{-1}R(K^{(r)})f(x) \\ &+ \tfrac{1}{4}h^4\mu_2(K)^2 f^{(r+2)}(x) + o\{(nh^{2r+1})^{-1} + h^4\}.\end{aligned} \quad (2.34)$$

It follows that the MSE-optimal bandwidth for estimating $f^{(r)}(x)$ is of order $n^{-1/(2r+5)}$. Therefore, estimation of $f'(x)$ requires a bandwidth of order $n^{-1/7}$ compared to the optimal $n^{-1/5}$ for estimation of f itself. Moreover, the optimal MSE and MISE is of order $n^{-4/(2r+5)}$. This rate becomes slower for higher values of r reflecting the increasing difficulty inherent in the problems of estimating higher derivatives (see e.g. Stone, 1982).

2.13 Bibliographical notes

2.1 An extensive study of the British household incomes data using kernel density estimation was carried out by Marron and Schmitz (1992). Park and Marron (1990) and Wand, Marron and Ruppert (1991) have also used kernel density estimation to analyse this data set.

2.2 The basic ideas of kernel density estimation first appeared in a technical report by E. Fix and J. L. Hodges (1951). This report was published out of historical interest as Fix and Hodges (1989). See also Silverman and Jones (1989). Akaike (1954) also contains some of these basic ideas. However, it was Rosenblatt (1956) and Parzen (1962) that provided the stimulus for considerable further interest in kernel methodology. Recent books on kernel density estimation include Silverman (1986), Härdle (1990a) and Scott (1992).

2.3 Detailed studies of the kernel density estimator with respect to the MIAE criterion, or expected L_1 norm, are given in Devroye and Györfi (1985) and Devroye (1987). Hall (1987) derived asymptotic theory for the kernel density estimator under Kullback-Leibler loss. Marron and Tsybakov (1993) proposed some interesting alternative error criteria.

2.4 For more on the material presented in Section 2.4 see, for example, Serfling (1980).

2.5 The asymptotic MSE and MISE calculations for the kernel density estimator were first performed by Rosenblatt (1956) and Parzen (1962). Since then there has been a great deal of theoretical investigation into the kernel density estimator. An important contribution is Farrell (1972) which was the first to derive lower bounds on the convergence rates of kernel estimators. For further theoretical results see, for example, Prakasa Rao (1983), Nadaraya (1989) and Rosenblatt (1991). The first MISE analyses of the histogram were performed by Scott (1979) and Diaconis and Freedman (1981).

2.6 Exact MSE and MISE calculations were first performed by Fryer (1976) and Deheuvels (1977a) for estimation of normal densities. Marron and Wand (1992) extended MISE calculations to the case of normal mixture densities. This reference also contains many uses of exact MISE calculations.

2.7 Watson and Leadbetter (1963) showed how to derive the kernel that is MISE-optimal for a given fixed density and finite sample size. Cline (1988) defined the notion of admissibility for kernels and showed that asymmetric and multimodal kernels are inadmissible. The canonical kernel idea is due to Marron and Nolan (1989). The argument that leads to the derivation of the asymptotically optimal second-order kernel was first given by Hodges and Lehmann (1956), although in a different context. Epanechnikov (1969) was the first to consider that problem in the density estimation context and give a comparison of common kernels in asymptotic performance terms. Granovsky and Müller (1991) is a recent reference on optimal kernel theory in more general contexts.

2.8 The idea of bias reduction using a higher-order kernel dates back to Parzen (1962) and Bartlett (1963). The sinc kernel density estimator, also called the *Fourier integral density estimator*, was first proposed by Davis (1975). Schucany and Sommers (1977) applied the generalised jackknife to bias reduction in kernel density estimation and showed that it is equivalent to using a higher-order kernel. Jones and Foster (1993) expanded on this theme. Optimal kernel theory for higher-order kernels can be found in Gasser, Müller and Mammitzsch (1985), Granovsky and Müller (1991) and Müller (1991), for example. A quantification of the practical gains that can be made using higher-order kernels is given by Marron and Wand (1992).

2.9 Results for optimizing $D(f)$ and other versions based on different scale measures are given in Terrell (1990). Wand and Devroye (1993) derived a scale-free measure of estimation difficulty based on the MIAE criterion.

2.10 Schucany (1989) discusses local bandwidth kernel density estimation. The nearest neighbour density estimator was first proposed by Loftsgaarden and Quesenberry (1965). Its mean squared error analysis is given in Mack and Rosenblatt (1979). See Silverman (1986) for several practical examples of the nearest neighbour density estimator.

The variable kernel density estimator was first proposed by Victor (1976) and Breiman, Meisel and Purcell (1977). Abramson (1982) derived the square-root law for achieving higher-order bias. Jones (1990) clarified the main differences between local and variable kernel approaches. Recent theoretical studies on variable kernel methodology include Hall (1990), Hall (1992) and Jones, McKay and Hu (1994).

Transformation kernel density estimation was first considered by Devroye and Györfi (1985) and Silverman (1986). Wand, Marron and Ruppert (1991) and Ruppert and Wand (1992) first considered choosing the transformation from a parametric family. Ruppert and Cline (1994) considered nonparametric estimation of the transformation.

2.11 An overview of boundary kernels can be obtained from Müller (1991) and Jones (1993). References for kernel density estimation at discontinuities include van Eeden (1987), Swanepoel (1987) and Cline and Hart (1991).

2.12 Bhattacharya (1967) and Singh (1979, 1987) made important contributions to kernel density derivative estimation.

2.14 Exercises

2.1 Derive the expression for $\text{Var}\{\hat{f}(x;h)\}$ given at (2.5).

2.2 Show that the formula for $\text{MISE}\{\hat{f}(\cdot;h)\}$ given at (2.8) follows from (2.7).

2.3 Determine the leading terms of the following sequences as $n \to \infty$:
(a) $3n^{4/5}/(2n^{1/3}+1)^5$,
(b) $(3+5/n)(4+1/n) - 2(6+7/n)$,
(c) $n^{-1} - (n^2+1)^{-1/2}$,
(d) $7^{1/n^2} - 2(7^{2/n^2}) + 7^{3/n^2}$,
(e) $n^{-1/10} - \tan^{-1}(n^{-1/10})$.

2.4 Use Taylor's theorem to expand the following sequences to terms of order h_n^2, where $h = h_n \to 0$ as $n \to \infty$:
(a) $\sin^2(\pi/4 + h)$,
(b) $e^{h^2+h} - e^{h^2}$,
(c) $\exp(-e^{x-h})$,
(d) $(x-h)/(x+h)$.

2.5 Using standard results concerning the moment generating function of the normal distribution, verify (2.9).

2.6 As described in Section 2.12, a kernel estimator for f', the first derivative of the density f, is

$$\hat{f}'(x;h) = (d/dx)\hat{f}(x;h) = n^{-1}h^{-2}\sum_{i=1}^{n} K'\{(x-X_i)/h\}$$

where K is a second-order kernel with a square integrable derivative.
(a) Assuming that f has three continuous derivatives and $h \to 0$, $nh^3 \to \infty$ as $n \to \infty$, show that the bias and variance of $\hat{f}'(x;h)$ satisfy

$$E\hat{f}'(x;h) - f'(x) = \tfrac{1}{2}h^2\mu_2(K)f'''(x) + o(h^2)$$

and

$$\text{Var}\{\hat{f}'(x;h)\} = n^{-1}h^{-3}R(K')f(x) + o(n^{-1}h^{-3}).$$

(b) Derive expressions for $\text{AMISE}\{\hat{f}'(\cdot;h)\}$, h_{AMISE} and

$$\inf_{h>0} \text{AMISE}\{\hat{f}'(\cdot;h)\}.$$

What is the rate of convergence of $\inf_{h>0} \text{MISE}\{\hat{f}'(\cdot;h)\}$?

2.7 Based on a sample $X_1, \ldots, X_n$, a *biweight M-estimate of location* is defined to be a solution, $\hat{\theta}_c$, to

$$\sum_{i=1}^{n} \chi_c\left(\frac{X_i - \theta}{\hat{\sigma}}\right) = 0$$

where $\chi_c(x) = x(c^2 - x^2)1_{\{|x|<c\}}$ and $\hat{\sigma}$ is an estimate of scale. Show that $\hat{\theta}_c$ is equal to a mode of the kernel density estimate based on $X_1, \ldots, X_n$ with bandwidth $h = c\hat{\sigma}$ and the biweight kernel $K(x) = \frac{15}{16}(1-x^2)^2 1_{\{|x|<1\}}$.

2.8 Let f_1 be the normal mixture density given by (2.3) and $\hat{f}_1(x;h)$ be a kernel estimator of f_1 based on the standard normal kernel and a random sample from f_1.
(a) Show that

$$E\hat{f}_1(x;h) = \tfrac{3}{4}\phi_{(1+h^2)^{1/2}}(x) + \tfrac{1}{4}\phi_{((1/3)^2+h^2)^{1/2}}(x - \tfrac{3}{2}).$$

(b) Verify that $\lim_{h \to 0} E\hat{f}_1(x;h) = f_1(x)$.

2.9 Let $K(x) = \tfrac{3}{4}(1-x^2)1_{\{|x|<1\}}$ be the Epanechnikov kernel and $f(x) = e^x e^{-e^x}$ be the extreme value density. Show that the AMISE-optimal bandwidth for estimation of f based on the kernel density estimator having kernel K is

$$h_{\text{AMISE}} = (60/n)^{1/5}.$$

2.10 Using (2.15), show that the MISE of the kernel estimator of the general mixture normal density using the standard normal kernel is given by (2.18).

2.11 Suppose that $X_1, \ldots, X_n$ are a random sample having density f. Let I be a random variable uniformly distributed on $\{1, \ldots, n\}$ and
$$Y = X_I + hZ$$
where Z has density K and is independent of $X_1, \ldots, X_n$ and I.
  (a) Show that, conditional on $X_1, \ldots, X_n$, Y has density $\hat{f}(\cdot; h)$, the kernel density estimate based on K.
  (b) Show that, conditional on $X_1, \ldots, X_n$, the variance of the density $\hat{f}(\cdot; h)$ exceeds the sample variance with divisor n by $h^2 R(K)$.

2.12 Show that (2.19) is satisfied if and only if $\delta = \delta_0$ where δ_0 is given by (2.20).

2.13
  (a) Derive the canonical kernel from the class $\{K_\delta : \delta > 0\}$ where K is the triangular kernel.
  (b) Verify the value of $\{C(K^*)/C(K)\}^{5/4}$ given in Table 2.1 for this kernel.

2.14 Show that $C(K)$ is invariant to the scale of K. That is, show that $C(K_{\delta_1}) = C(K_{\delta_2})$ for any $\delta_1, \delta_2 > 0$.

2.15
  (a) Use (2.23) to obtain a sixth-order kernel from (2.24).
  (b) Likewise, obtain a fourth-order kernel from the biweight kernel.
  (c) Derive an eighth-order kernel from (a).

2.16
  (a) Prove that the kernel $G_{[k]}$ given by (2.26) is a kth-order kernel.
  (b) Show that the answer to Exercise 2.15(a) corresponds to $G_{[6]}$ as given by (2.26).

2.17 Let $\varphi_g(t) = \int e^{itx} g(x)\, dx$ denote the Fourier transform of the real-valued function g. If g is a density then $\varphi_g(t)$ is often called the *characteristic function* of g.
  (a) Let f and g be two real-valued functions. Show that
$$\varphi_{f*g}(t) = \varphi_f(t)\varphi_g(t).$$
and that
$$\int f(x)g(x)\,dx = (2\pi)^{-1} \int \varphi_f(t)\overline{\varphi_g(t)}\,dt$$

where $\bar{z}$ denotes the conjugate of a complex number z. The second result is called *Parseval's identity*.
(b) Suppose that K is symmetric. Apply the results from (a) to (2.8) to show that

$$\text{MISE}\{\hat{f}(\cdot;h)\} = (2\pi nh)^{-1} \int \kappa(t)^2 \, dt$$
$$+ (2\pi)^{-1} \int \{(1-n^{-1})\kappa(ht)^2 - 2\kappa(ht) + 1\} |\varphi_f(t)|^2 \, dt$$

where $\kappa(t) = \int e^{itx} K(x) \, dx$.

2.18 The characteristic function of the sinc kernel is given by

$$\int e^{itx} \sin x / (\pi x) \, dx = 1_{\{|t| \le 1\}}.$$

Using this result and Exercise 2.17, show that the MISE of the sinc kernel density estimator is given by

$$\text{MISE}\{\hat{f}(\cdot;h)\} = (\pi nh)^{-1} - \pi^{-1}(1+n^{-1}) \int_0^{1/h} |\varphi_f(t)|^2 \, dt$$
$$+ \int f(x)^2 \, dx$$

and the MISE-optimal bandwidth satisfies

$$|\varphi_f(1/h_{\text{MISE}})|^2 = (n+1)^{-1}$$

provided $|\varphi_f(t)| > 0$ for all t (Davis, 1975).

2.19 The *Laplace kernel* $K(x) = \frac{1}{2}e^{-|x|}$ has characteristic function $\varphi_K(t) = (1+t^2)^{-1}$ while the exponential density, $f(x) = e^{-x}1_{\{x>0\}}$, has characteristic function $\varphi_f(t) = (1-it)^{-1}$. Use these results and Exercise 2.17 to derive an expression for the MISE of the kernel estimator of the exponential density using the Laplace kernel.

2.20 For the sinc kernel estimator use the second result of Exercise 2.18 to derive expressions for h_{MISE} when f is
(a) the standard normal density.
(b) the exponential density $f(x) = e^{-x}1_{\{x>0\}}$.
(c) the Cauchy density $f(x) = \{\pi(1+x^2)\}^{-1}$.

For the Cauchy density obtain an expression for
$$\inf_{h>0} \text{MISE}\{\hat{f}(\cdot;h)\}.$$
What is the rate of convergence of the minimum MISE?

2.21 Verify the result for $D(f^*)/D(f)$ in the case where f is the gamma(3) density.

2.22 Hölder's inequality states that for two real-valued integrable functions α and β and positive integers p and q satisfying $1/p + 1/q = 1$ we have
$$\int \alpha(x)\beta(x)\,dx \leq \left\{\int \alpha(x)^p\,dx\right\}^{1/p} \left\{\int \beta(x)^q\,dx\right\}^{1/q}.$$
Use this to prove (2.30).

2.23 Let $\hat{f}_T(\cdot;h,t)$ be the transformation kernel density estimator and g be the density of $t(X)$.

(a) Show that the bias of $\hat{f}_T(x;h,t)$ is given by
$$E\{\hat{f}_T(x;h,t)\} - f(x) = \tfrac{1}{2}h^2\mu_2(K)t'(x)g''(t(x)) + o(h^2).$$

(b) Show that the variance of $\hat{f}_T(x;h,t)$ is given by
$$\text{Var}\{\hat{f}_T(x;h,t)\} = (nh)^{-1}R(K)t'(x)^2 g(t(x)) + o\{(nh)^{-1}\}.$$

(c) Combine these to show that the AMISE of $\hat{f}_T(\cdot;h,t)$ is given by
$$\text{AMISE}\{\hat{f}_T(\cdot;h,t)\} = (nh)^{-1}R(K)E\{t'(X)\}$$
$$+ \tfrac{1}{4}h^4\mu_2(K)^2 \int t'\{t^{-1}(x)\}g''(x)^2 dx.$$

2.24 If K is a second-order kernel, show that
$$K^\alpha(x) = \Big([2\alpha\{1-\nu_{0,\alpha}(K)\} + \nu_{1,\alpha}(K)]\nu_{0,\alpha}(K)$$
$$- \nu_{1,\alpha}(K)\{1-\nu_{0,\alpha}(K)\}\Big)^{-1}$$
$$\times \Big([2\alpha\{1-\nu_{0,\alpha}(K)\} + \nu_{1,\alpha}(K)]K(x)$$
$$- \nu_{1,\alpha}(K)K(2\alpha - x)\Big)$$

2.14 EXERCISES

is a boundary kernel. That is, show that

$$\int_{-1}^{\alpha} K^{\alpha}(x)\,dx = 1 \quad \text{and} \quad \int_{-1}^{\alpha} xK^{\alpha}(x)\,dx = 0.$$

2.25 Show that $K_h^{(r)} * f = K_h * f^{(r)}$. Hence verify the mean squared error expression for $\hat{f}^{(r)}(x;h)$ given at (2.34).

CHAPTER 3

Bandwidth selection

3.1 Introduction

The practical implementation of the kernel density estimator requires the specification of the bandwidth h. This choice is very important as was shown graphically in Section 2.2 and through the AMISE analysis in Section 2.5. There are many situations where it is satisfactory to choose the bandwidth subjectively by eye. This would involve looking at several density estimates over a range of bandwidths and selecting the density that is the "most pleasing" in some sense. One such strategy is to begin with a large bandwidth and to decrease the amount of smoothing until fluctuations that are more "random" than "structural" start to appear. This approach is more viable when the user has reasons to believe that there is certain structure in the data, such as knowledge of the position of modes. However, there are also many circumstances where it is very beneficial to have the bandwidth automatically selected from the data. One reason is that it can be very time consuming to select the bandwidth by eye if there are many density estimates required for a given problem. Another is that, in many cases, the user has no prior knowledge about the structure of the data and would not have any feeling for which bandwidth gives an estimate closest to the true density. When kernel estimators are used as components of larger statistical procedures, automatic bandwidth selection is usually necessary.

A method that uses the data $X_1, \ldots, X_n$ to produce a bandwidth $\hat{h}$ is called a *bandwidth selector*. The bandwidth selection problem is present in all types of kernel estimation, including the scatterplot smoothing problem dealt with in Chapter 5, although kernel density estimation provides a convenient setting for the development of many of the key ideas.

3.1 INTRODUCTION

Currently available bandwidth selectors can be roughly divided into two classes. The first class consists of simple easily computable formulae which aim to find a bandwidth that is "reasonable" for a wide range of situations, but without out any mathematical guarantees of being close to the optimal bandwidth. We will call such bandwidth selectors *quick and simple* and discuss some proposals in Section 3.2. Quick and simple bandwidth selectors are motivated by the need to have fast automatically generated kernel estimates for algorithms that require many curve estimation steps as well as providing a reasonable starting point for subjective choice of the smoothing parameter.

The second type of bandwidth selector will be labelled as *hi-tech* since such selection procedures are based on more involved mathematical arguments and require considerably more computational effort, but aim to give a good answer for very general classes of underlying functions. Each of the hi-tech bandwidth selectors that we discuss in this chapter can be motivated through aiming to minimise MISE$\{\hat{f}(\cdot;h)\}$ and can be shown to attain this goal *asymptotically* to some extent. Such a bandwidth selector is said to be *consistent* with respect to MISE. This chapter is devoted to a presentation and comparison of such MISE-driven bandwidth selectors. It should be pointed out that there exist approaches to bandwidth selection based on other loss criteria. However, since their analysis is more difficult we will restrict attention to the simpler MISE-driven selectors.

At the time of writing the field of bandwidth selection remains fairly unsettled, with new selectors being developed and several unresolved issues. Despite this, there has been a considerable amount of path-breaking research on the problem in recent years.

This chapter aims to reach a compromise between simple presentation of the main ideas and representative coverage of current approaches to practical bandwidth selection.

Throughout this section we will assume that $\hat{f}(\cdot;h)$ is a kernel density estimator based on a random sample $X_1, \ldots, X_n$ having density f. Furthermore, we will assume that $\hat{f}(\cdot;h)$ uses a second-order kernel K and that f is sufficiently well-behaved for all arguments involving differentiability and integrability assumptions to be valid.

3.2 Quick and simple bandwidth selectors

In this section we describe two commonly used quick and simple ideas for selecting the bandwidth of the kernel density estimator $\hat{f}(\cdot;h)$. Quick and simple rules also play an important role in the implementation of several hi-tech bandwidth selectors, as we will see further on.

3.2.1 Normal scale rules

A *normal scale* bandwidth selector simply involves using the bandwidth that is AMISE-optimal for the normal density having the same scale as that estimated for the underlying density. As shown in Section 2.5, the bandwidth that minimises MISE$\{\hat{f}(\cdot;h)\}$ asymptotically is

$$h_{\text{AMISE}} = \left[\frac{R(K)}{\mu_2(K)^2 R(f'')n}\right]^{1/5}.$$

If f is normal with variance σ^2 then it is easily shown (Exercise 3.1) that

$$h_{\text{AMISE}} = \left[\frac{8\pi^{1/2} R(K)}{3\mu_2(K)^2 n}\right]^{1/5} \sigma. \qquad (3.1)$$

A normal scale bandwidth selector is obtained from (3.1) by simply replacing σ by $\hat{\sigma}$:

$$\hat{h}_{\text{NS}} = \left[\frac{8\pi^{1/2} R(K)}{3\mu_2(K)^2 n}\right]^{1/5} \hat{\sigma} \qquad (3.2)$$

where $\hat{\sigma}$ is some estimate of σ (e.g. Silverman, 1986, pp.45–47). Common choices of $\hat{\sigma}$ are the sample standard deviation s and the standardised interquartile range

$$\hat{\sigma}_{\text{IQR}} = (\text{sample interquartile range})/\{\Phi^{-1}(\tfrac{3}{4}) - \Phi^{-1}(\tfrac{1}{4})\}$$

where Φ^{-1} is the standard normal quantile function. Note that the normalising factor in the denominator of $\hat{\sigma}_{\text{IQR}}$ is the population interquartile range of the standard normal density and is approximately equal to 1.349. Use of $\hat{\sigma}_{\text{IQR}}$ guards against outliers if f has heavy tails. It is sometimes recommended that the smaller of s and $\hat{\sigma}_{\text{IQR}}$ be used (Silverman, 1986, p.47) to lessen the chances of oversmoothing. More sophisticated scale estimates have also

3.2 QUICK AND SIMPLE BANDWIDTH SELECTORS

been studied and recommended (Janssen, Marron, Veraverbeke and Sarle, 1995).

Normal scale bandwidth selectors provide a quick "first guess" bandwidth and can be expected to give reasonable answers when the data are close to normal. However, for departures from normality such as multimodality, which one usually hopes to be detected by a density estimate, normal scale bandwidth selectors tend to oversmooth and mask important features in the data.

3.2.2 Oversmoothed bandwidth selection rules

The *oversmoothing* or *maximal smoothing* principle relies on the fact that there is a simple upper bound for the AMISE-optimal bandwidth for estimation of densities with a fixed value of a particular scale measure. For example, it can be shown (Terrell, 1990, Theorem 1) that

$$h_{\text{AMISE}} \leq \left[\frac{243R(K)}{35\mu_2(K)^2 n}\right]^{1/5} \sigma \qquad (3.3)$$

for all densities having standard deviation σ and that this bound is attained by the beta(4,4) or triweight density, pictured in Figure 2.11. Similar results can be shown to hold for other scale measures (Terrell and Scott, 1985, Terrell, 1990). The above bound on h_{AMISE} motivates the *oversmoothed* bandwidth selector

$$\hat{h}_{\text{OS}} = \left[\frac{243R(K)}{35\mu_2(K)^2 n}\right]^{1/5} s$$

where s is the sample standard deviation. It is also possible to base $\hat{h}_{\text{OS}}$ on other common scale measures (Terrell, 1990). While $\hat{h}_{\text{OS}}$ will give too large a bandwidth for optimal estimation of a general density f it provides an excellent starting point for subjective choice of the bandwidth. A sensible graphical strategy is to plot an estimate with bandwidth $\hat{h}_{\text{OS}}$ and then successively look at plots based on convenient fractions of $\hat{h}_{\text{OS}}$ to see what features are present in the data for various amount of smoothing.

Figure 3.1 illustrates this idea for the *Old Faithful* data set, consisting of 107 eruption times in minutes for the Old Faithful Geyser in Yellowstone National Park (source: Silverman, 1986, p.8).

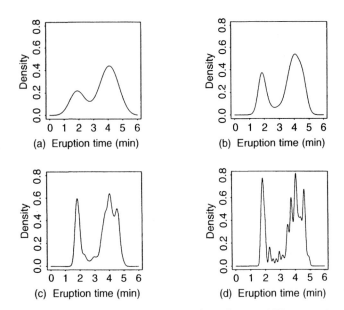

Figure 3.1. *Density estimates based on 107 eruption times (minutes) of the Old Faithful Geyser. The normal kernel is used. Bandwidths are (a)* $\hat{h}_{OS}$, *(b)* $\hat{h}_{OS}/2$, *(c)* $\hat{h}_{OS}/4$ *and (d)* $\hat{h}_{OS}/8$.

Figure 3.1 (a) shows the density estimate based on the normal kernel with bandwidth $\hat{h}_{OS} = 0.467$. An important point to note from this estimate is that it is bimodal, despite the fact that it is oversmoothed. This provides very strong evidence in favour of eruption times exhibiting a bimodal distribution, with distinct clusters centred around about 1.9 minutes and 4.2 minutes. Decreasing the bandwidth by a factor of 2 results in Figure 3.1 (b). This density estimate retains the bimodal structure of the oversmoothed estimate, but resolves these features more sharply. Halving the bandwidth again leads to Figure 3.1 (c). In this case five modes are present, the smallest three almost certainly an artifact of having too small a bandwidth. Figure 3.1 (d), with bandwidth $\hat{h}_{OS}/8$, leads to a very undersmoothed estimate which is far too wiggly to be a serious contender for modelling eruption times. Of these four estimates, (b) is the most pleasing since it appears to reach a good compromise between highlighting features in the data and containing its variability.

The oversmoothed and normal scale bandwidth selectors based on standard deviation are closely related in the sense that

$$\hat{h}_{NS}/\hat{h}_{OS} = (280\pi^{1/2}/729)^{1/5} \simeq 0.93.$$

3.3 Least squares cross-validation

This is because the normal density is close to obtaining the upper bound in (3.3).

3.3 Least squares cross-validation

We will now begin our description of a selection of hi-tech bandwidth selectors. Among the earliest fully automatic and consistent bandwidth selectors were those based on cross-validation ideas. *Least squares cross-validation* (LSCV) (Rudemo, 1982, Bowman, 1984) is the name given to a conceptually simple and appealing bandwidth selector. Its motivation comes from expanding the MISE of $\hat{f}(\cdot;h)$ to obtain

$$\text{MISE}\{\hat{f}(\cdot;h)\} = E\int \hat{f}(x;h)^2\,dx - 2E\int \hat{f}(x;h)f(x)\,dx + \int f(x)^2\,dx.$$

Notice that the $\int f(x)^2\,dx$ term does not depend on h, so minimization of $\text{MISE}\{\hat{f}(\cdot;h)\}$ is equivalent to minimization of

$$\text{MISE}\{\hat{f}(\cdot;h)\} - \int f(x)^2\,dx = E\left[\int \hat{f}(x;h)^2\,dx - 2\int \hat{f}(x;h)f(x)\,dx\right].$$

The right-hand side is unknown since it depends on f. However, it can be shown (Exercise 3.3) that an unbiased estimator for this quantity is

$$\text{LSCV}(h) = \int \hat{f}(x;h)^2\,dx - 2n^{-1}\sum_{i=1}^{n}\hat{f}_{-i}(X_i;h)$$

where

$$\hat{f}_{-i}(x;h) = (n-1)^{-1}\sum_{j\neq i}^{n} K_h(x-X_j)$$

is the density estimate based on the sample with X_i deleted, often called the "leave-one-out" density estimator. This is the reason for the term "cross-validation" which refers to the use of part of a sample to obtain information about another part. It therefore seems reasonable to choose h to minimise LSCV(h). We denote the bandwidth chosen according to this strategy by $\hat{h}_{\mathrm{LSCV}}$. It is sometimes the case that LSCV(h) has more than one local minimum (Hall and Marron, 1991a).

Figure 3.2 shows LSCV(h) versus $\log_{10}(h)$ for two particular samples of size $n = 100$ from the standard normal density using the normal kernel.

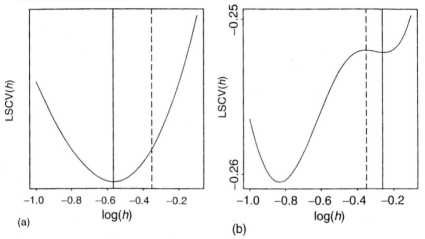

Figure 3.2. *Examples of* LSCV(h) *for two samples of* 100 $N(0,1)$ *observations. A* $\log_{10}$ *scale is used on the horizontal axis. The dashed vertical line shows the position of* $\log_{10}(h_{\mathrm{MISE}})$. *The solid vertical lines show the position of* $\log_{10}(\hat{h}_{\mathrm{LSCV}})$ *if* $\hat{h}_{\mathrm{LSCV}}$ *is taken to correspond to the largest local minimum. The kernel is the standard normal density.*

Figure 3.2 (b) is an example of LSCV having two minima. The $\log_{10}$ of the MISE-optimal bandwidth $h_{\mathrm{MISE}} \simeq 0.445$ is shown by the dashed vertical line. Notice that the actual minimum is much smaller than h_{MISE}, while the larger minimiser is considerably closer to h_{MISE}. This phenomenon has led to the suggestion that $\hat{h}_{\mathrm{LSCV}}$ be taken to correspond to the *largest* local minimiser of LSCV(h) (Marron, 1993). The multiple minima phenomenon also means that care needs to be taken when finding $\hat{h}_{\mathrm{LSCV}}$ in practice.

Studies have shown (e.g. Hall and Marron, 1987a, Park and Marron, 1990) that the theoretical and practical performance of this bandwidth selector are somewhat disappointing. In particular, $\hat{h}_{\mathrm{LSCV}}$ is highly variable (see Figure 3.3). This has since led to the

3.4 Biased cross-validation

Instead of the exact MISE formula used by least squares cross-validation, *biased cross-validation* (BCV) (Scott and Terrell, 1987) is based on the formula for the asymptotic MISE:

$$\text{AMISE}\{\hat{f}(\cdot;h)\} = (nh)^{-1}R(K) + \tfrac{1}{4}h^4\mu_2(K)^2 R(f''). \quad (3.4)$$

The BCV objective function is obtained by replacing the unknown $R(f'')$ in (3.4) by the estimator

$$\widetilde{R(f'')} = R(\hat{f}''(\cdot;h)) - (nh^5)^{-1}R(K'')$$
$$= n^{-2}\sum\sum_{i\neq j}(K_h'' * K_h'')(X_i - X_j)$$

to give

$$\text{BCV}(h) = (nh)^{-1}R(K) + \tfrac{1}{4}h^4\mu_2(K)^2 \widetilde{R(f'')}.$$

The BCV bandwidth selector, which we denote by $\hat{h}_{\text{BCV}}$, is the minimiser of $\text{BCV}(h)$. This selector is really a hybrid of cross-validation and "plug-in" bandwidth selection, as described in Section 3.6, since it involves replacement of the unknown $R(f'')$ by the cross-validatory kernel estimator $\widetilde{R(f'')}$.

The main attraction of $\hat{h}_{\text{BCV}}$ is that it is more stable than $\hat{h}_{\text{LSCV}}$, in the sense that its asymptotic variance is considerably lower (see Section 3.8). However, this reduction in variance comes at the price of an increase in bias, with $\hat{h}_{\text{BCV}}$ tending to be larger than the MISE-optimal bandwidth. This is illustrated in Figure 3.3, which shows kernel density estimates based on $\hat{h}_{\text{LSCV}}$ and $\hat{h}_{\text{BCV}}$ bandwidths obtained from 500 simulated samples of size $n = 100$ from the normal mixture density f_1 defined at (2.3). The estimates are on a $\log_{10}$ scale with $\log_{10}(h_{\text{MISE}})$ subtracted. The bandwidths for these estimates were obtained using the normal scale rule based on the sample standard deviation. This is a reasonable choice since $\hat{h}_{\text{LSCV}}$ and $\hat{h}_{\text{BCV}}$ each have asymptotically normal distributions (see Section 3.8). The vertical line at 0 shows the position of h_{MISE}. The normal kernel is used throughout.

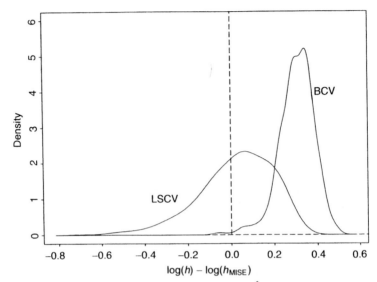

Figure 3.3. *Density estimates of* $\log_{10}(\hat{h}_{\text{LSCV}}) - \log_{10}(h_{\text{MISE}})$ *and* $\log_{10}(\hat{h}_{\text{BCV}}) - \log_{10}(h_{\text{MISE}})$. *Selected bandwidths are based on 500 simulated samples of size* $n = 100$ *from the normal mixture density* f_1 *defined at (2.3).*

The relative stability of BCV is manifest through the tightness of the distribution of the density of the $\hat{h}_{\text{BCV}}$'s compared to those of the $\hat{h}_{\text{LSCV}}$'s. However, the fact that the BCV distribution is situated to the right of 0 indicates the positive bias present in the $\hat{h}_{\text{BCV}}$'s, while there is no noticeable bias in the $\hat{h}_{\text{LSCV}}$'s. This variance-bias trade-off for bandwidth selectors is also present for other types of rules and is mathematically quantified in Section 3.8.

Like LSCV, the BCV criterion function occasionally has more than one local minimum, as well as being globally minimised at $h = 0$. Rules for choosing among local minima have been suggested in the literature (Scott, 1992, p.167, Marron, 1993).

3.5 Estimation of density functionals

An important component of many of the current hi-tech univariate bandwidth selectors is the estimation of integrated squared density derivatives. This is because they arise in various expressions for optimal bandwidths. The general integrated squared density derivative functional is

$$R(f^{(s)}) = \int f^{(s)}(x)^2 \, dx.$$

However, it is a simple exercise in using integration by parts to show that, under sufficient smoothness assumptions on f,

$$R(f^{(s)}) = (-1)^s \int f^{(2s)}(x) f(x) \, dx.$$

It is therefore sufficient to study estimation of functionals of the form

$$\psi_r = \int f^{(r)}(x) f(x) \, dx$$

for r even. Note that the sign of ψ_{2s} is the same as that of $(-1)^s$ and $\psi_r = 0$ if r is odd. We prefer the ψ_r notation to the usual $R(f^{(r)})$ notation because its extension to the multivariate context is more straightforward.

Note that

$$\psi_r = E\{f^{(r)}(X)\}.$$

This motivates the estimator

$$\hat{\psi}_r(g) = n^{-1} \sum_{i=1}^{n} \hat{f}^{(r)}(X_i; g) = n^{-2} \sum_{i=1}^{n} \sum_{j=1}^{n} L_g^{(r)}(X_i - X_j)$$

(Hall and Marron, 1987b, Jones and Sheather, 1991) where g and L are, respectively, a bandwidth and kernel that are possibly different from h and K.

The asymptotic MSE properties of $\hat{\psi}_r$ are of fundamental importance for the bandwidth selectors described in the following two sections. We will give their derivation here under the following assumptions:

(i) the kernel L is a symmetric kernel of order k, $k = 2, 4, \ldots$, possessing r derivatives, such that

$$(-1)^{(r+k)/2+1} L^{(r)}(0) \mu_k(L) > 0.$$

(ii) the density f has p continuous derivatives that are each ultimately monotone, where $p > k$.

(iii) $g = g_n$ is a positive-valued sequence of bandwidths satisfying

$$\lim_{n\to\infty} g = 0 \quad \text{and} \quad \lim_{n\to\infty} ng^{2r+1} = \infty.$$

Condition (i) is satisfied by most of the kernels that are used in practice.

We seek a large sample approximation to

$$\text{MSE}\{\hat{\psi}_r(g)\} = E\{\hat{\psi}_r(g) - \psi_r\}^2.$$

First note that

$$\hat{\psi}_r(g) = n^{-1} L_g^{(r)}(0) + n^{-2} \sum\sum\nolimits_{i\neq j} L_g^{(r)}(X_i - X_j),$$

the first term being independent of the data. Clearly,

$$E\hat{\psi}_r(g) = n^{-1} L_g^{(r)}(0) + (1 - n^{-1}) E\{L_g^{(r)}(X_1 - X_2)\}$$

and, by Taylor's theorem and the smoothness assumptions on f,

$$E\{L_g^{(r)}(X_1 - X_2)\} = \int\int L_g^{(r)}(x - y) f(x) f(y) \, dx \, dy$$

$$= \int\int L_g(x - y) f(x) f^{(r)}(y) \, dx \, dy$$

$$= \int\int L(u) f(y + gu) f^{(r)}(y) \, du \, dy$$

$$= \int\int L(u) f^{(r)}(y)$$

$$\times \left\{ \sum_{\ell=0}^{k} (\ell!)^{-1} (ug)^\ell f^{(\ell)}(y) + O(g^{k+1}) \right\} du\, dy$$

$$= \psi_r + (k!)^{-1} \mu_k(L) g^k \psi_{r+k} + O(g^{k+2}).$$

The bias can therefore be expressed as

$$E\hat{\psi}_r(g) - \psi_r = n^{-1} g^{-r-1} L^{(r)}(0)$$
$$+ (k!)^{-1} g^k \mu_k(L) \psi_{r+k} + O(g^{k+2}).$$

3.5 ESTIMATION OF DENSITY FUNCTIONALS

It follows from Exercise 3.4 and the symmetry of $L^{(r)}$ for r even that

$$\text{Var}\{\hat\psi_r(g)\} = 2n^{-3}(n-1)\text{Var}\{L_g^{(r)}(X_1 - X_2)\} \tag{3.5}$$
$$+ 4n^{-3}(n-1)(n-2)\text{Cov}\{L_g^{(r)}(X_1 - X_2), L_g^{(r)}(X_2 - X_3)\}.$$

We will treat each component of the variance and covariance in turn. Firstly,

$$E[\{L_g^{(r)}(X_1 - X_2)\}^2] = \int\int \{L_g^{(r)}(x-y)\}^2 f(x)f(y)\,dx\,dy$$
$$= g^{-2r-1}\int\int L^{(r)}(u)^2 f(y+gu)f(y)\,du\,dy$$
$$= g^{-2r-1}\psi_0 R(L^{(r)}) + o(g^{-2r-1})$$

while

$$E\{L_g^{(r)}(X_1 - X_2)L_g^{(r)}(X_2 - X_3)\}$$
$$= \int\int\int L_g^{(r)}(x-y)L_g^{(r)}(y-z)f(x)f(y)f(z)\,dx\,dy\,dz$$
$$= \int\int\int L_g(x-y)L_g(y-z)f^{(r)}(x)f(y)f^{(r)}(z)\,dx\,dy\,dz$$
$$= \int\int\int L(u)L(v)f^{(r)}(y+gu)f(y)f^{(r)}(y-gv)\,du\,dv\,dy$$
$$= \int f^{(r)}(y)^2 f(y)\,dy + o(1).$$

Lastly, from above we have

$$E\{L_g^{(r)}(X_1 - X_2)\} = \psi_r + o(1).$$

Combining each of these approximations with (3.5) leads to

$$\text{Var}\{\psi_r(g)\} = 2n^{-2}g^{-2r-1}\psi_0 R(L^{(r)})$$
$$+ 4n^{-1}\left\{\int f^{(r)}(x)^2 f(x)\,dx - \psi_r^2\right\} + o(n^{-2}g^{-2r-1} + n^{-1}).$$

The asymptotic MSE is therefore

$$\text{MSE}\{\hat\psi_r(g)\} =$$
$$\{n^{-1}g^{-r-1}L^{(r)}(0) + (k!)^{-1}g^k \mu_k(L)\psi_{r+k}\}^2$$
$$+ 2n^{-2}g^{-2r-1}R(L^{(r)})\psi_0 + 4n^{-1}\left\{\int f^{(r)}(x)^2 f(x)\,dx - \psi_r^2\right\}$$
$$+ O(g^{2k+2}) + o(n^{-2}g^{-2r-1} + n^{-1}).$$

Notice that, because of our assumption about the sign of $L^{(r)}(0)\mu_k(L)$, the main bias term can be made to vanish by choosing g to equal

$$g_{\text{AMSE}} = \left[\frac{k!L^{(r)}(0)}{-\mu_k(L)\psi_{r+k}n}\right]^{1/(r+k+1)}. \tag{3.6}$$

While this choice reduces the squared bias of the MSE to be of order $n^{-(2k+4)/(r+k+1)}$ we also need to check the orders of the variance terms. Since $g_{\text{AMSE}} = O(n^{-1/(r+k+1)})$ we obtain the leading variance terms to be of orders $n^{-(2k+1)/(r+k+1)}$ and n^{-1} respectively. The first of these variance terms dominates the remaining squared bias term so it is clear that the rate of convergence of the minimum MSE depends only on the leading variance terms. It is easy to check that for $k < r$,

$$\inf_{g>0} \text{MSE}\{\hat{\psi}_r(g)\} \sim$$

$$2R(L^{(r)})\psi_0 \left[\frac{\mu_k(L)\psi_{r+k}}{-L^{(r)}(0)k!}\right]^{(2r+1)/(r+k+1)} n^{-(2k+1)/(r+k+1)}$$

while for $k > r$,

$$\inf_{g>0} \text{MSE}\{\hat{\psi}_r(g)\} \sim 4\left[\text{Var}\{f^{(r)}(X)\}\right]n^{-1}.$$

If $k = r$ then the two leading terms in the above expression are of the same order and the leading term of the minimum mean squared error is the sum of these terms. Therefore, the parametric rate of convergence of order n^{-1} is achievable, provided the kernel is at least of order r. Of course, choosing k to be higher means that p, the number of continuous derivatives that we assume for f, is higher as well.

Finally, we point out that the computation of $\hat{\psi}_r(g)$ can be very expensive if a direct algorithm is used. This is because it involves $O(n^2)$ operations. In practice it is recommended that binned approximations, as described in Appendix D, be used instead.

3.6 Plug-in bandwidth selection

3.6.1 Direct plug-in rules

Plug-in bandwidth selectors are based on the simple idea of "plugging in" estimates of the unknown quantities that appear in formulae for the asymptotically optimal bandwidth. In terms of the ψ_r functionals studied in the previous section the AMISE-optimal bandwidth is

$$h_{\text{AMISE}} = \left[\frac{R(K)}{\mu_2(K)^2 \psi_4 n}\right]^{1/5}.$$

Replacement of ψ_4 by the kernel estimator $\hat{\psi}_4(g)$ leads to the *direct plug-in* (DPI) rule

$$\hat{h}_{\text{DPI}} = \left[\frac{R(K)}{\mu_2(K)^2 \hat{\psi}_4(g) n}\right]^{1/5}.$$

Unfortunately, this rule is not fully automatic since $\hat{h}_{\text{DPI}}$ depends on the choice of the *pilot bandwidth* g. One way of choosing g is to appeal to the formula for the AMSE-optimal bandwidth for estimation of $\hat{\psi}_4(g)$. If the same second-order kernel K is used in $\hat{\psi}_4(g)$ then from (3.6) the AMSE-optimal bandwidth is

$$g_{\text{AMSE}} = \left[\frac{2K^{(4)}(0)}{-\mu_2(K)\psi_6 n}\right]^{1/7}.$$

However, this rule for choosing g has the same defect as $\hat{h}_{\text{DPI}}$ above: it depends on an unknown density functional, namely ψ_6. We could estimate ψ_6 using another kernel estimate, but its optimal bandwidth depends on ψ_8. This problem will not go away since it is apparent from (3.6) that the optimal bandwidth for estimating ψ_r depends on ψ_{r+2}.

The usual strategy for overcoming this problem is to estimate a ψ_r functional with a quick and simple estimate, such as a version of the normal scale rule described in Section 3.2.1. This means that we really have a family of direct plug-in bandwidth selectors that depend on the number of stages of functional estimation before a quick and simple estimate is used. Suppose that a direct plug-in rule involves ℓ successive kernel functional estimations, with the initial bandwidth chosen via a quick and simple procedure. We will call such a rule an *ℓ-stage direct plug-in* bandwidth selector and denote it by $\hat{h}_{\text{DPI},\ell}$. Note that the normal scale rule (3.2) can be thought of as being a zero-stage direct plug-in bandwidth selector.

Appendix C contains results that are very useful for computing quantities required for certain bandwidth selection strategies, especially those that involve normal scale rules and normal kernels. The following result is particularly useful and follows from Fact C.1.12 and Fact C.1.6 in Appendix C. If f is a normal density with variance σ^2 then, for r even,

$$\psi_r = \frac{(-1)^{r/2} r!}{(2\sigma)^{r+1}(r/2)!\pi^{1/2}}. \qquad (3.7)$$

EXAMPLE. A version of the two-stage plug-in bandwidth selector follows (Sheather and Jones, 1991). Note that we are using $L = K$, where K is a second-order kernel.

Step 1 Estimate ψ_8 using the normal scale estimate $\hat{\psi}_8^{\text{NS}} = 105/(32\pi^{1/2}\hat{\sigma}^9)$ where $\hat{\sigma}$ is an estimate of scale. The formula for $\hat{\psi}_8^{\text{NS}}$ is obtained from (3.7).

Step 2 Estimate ψ_6 using the kernel estimator $\hat{\psi}_6(g_1)$ where $g_1 = [-2K^{(6)}(0)/\{\mu_2(K)\hat{\psi}_8^{\text{NS}} n\}]^{1/9}$.

Step 3 Estimate ψ_4 using the kernel estimator $\hat{\psi}_4(g_2)$ where $g_2 = [-2K^{(4)}(0)/\{\mu_2(K)\hat{\psi}_6(g_1) n\}]^{1/7}$.

Step 4 The selected bandwidth is

$$\hat{h}_{\text{DPI},2} = [R(K)/\{\mu_2(K)^2 \hat{\psi}_4(g_2) n\}]^{1/5}.$$

■

Another selection problem should now be apparent: how should one choose the value of ℓ, the number of stages of functional estimation? To give an idea of the effect of ℓ on the performance of $\hat{h}_{\text{DPI},\ell}$, density estimates of 500 simulated bandwidths (on a $\log_{10}$ scale with $\log_{10} h_{\text{MISE}}$ subtracted) are plotted in Figure 3.4 for the direct plug-in rule with $\ell = 0, 1, 2$ and 3 and sample standard deviation used in the normal scale rule for the initial bandwidth.

3.6 PLUG-IN BANDWIDTH SELECTION

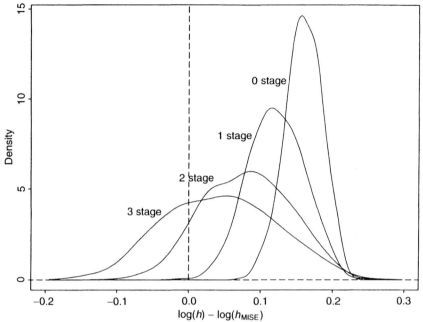

Figure 3.4. *Density estimates based on values of* $\log_{10}(\hat{h}_{\mathrm{DPI},\ell}) - \log_{10}(h_{\mathrm{MISE}})$ *for* $\ell = 0, 1, 2, 3$. *Selected bandwidths are based on 500 simulated samples of size* $n = 100$ *from the normal mixture density* f_1 *defined at (2.3)*.

The selected bandwidths were obtained from samples of size $n = 100$ from the normal mixture density f_1 defined at (2.3). The vertical line shows the position of h_{MISE}. The estimated density of the zero-stage bandwidth $\hat{h}_{\mathrm{DPI},0}$, which we have also denoted by $\hat{h}_{\mathrm{NS}}$, is represented by the right-most curve. Since these bandwidths are based on a normal distribution assumption it is not surprising that they tend to be larger than h_{MISE} (since the optimal bandwidth for normal data is also larger). Also note the relative tightness of the distribution of the $\hat{h}_{\mathrm{DPI},0}$. This is because the variability between different bandwidths is due only to different standard deviation estimates. As ℓ increases we see that the bandwidth selector becomes less biased, since the dependence on the normal scale rule diminishes. However, the extra functional estimation steps for larger ℓ lead to the selector being more variable. This stage selection problem is also present in related hi-tech rules, such as smoothed cross-validation described in Section 3.7. At the time of writing, there was no method for objective choice of ℓ. However, theoretical considerations (Aldershof, 1991; Park and

Marron, 1992) favour taking ℓ to be at least 2, with $\ell = 2$ being a common choice.

3.6.2 Solve-the-equation rules

Also motivated by the formula for the AMISE-optimal bandwidth, *solve-the-equation* (STE) rules (Scott, Tapia and Thompson, 1977, Sheather, 1986, Park and Marron, 1990, Sheather and Jones, 1991, Engel, Herrmann and Gasser, 1995) require that h be chosen to satisfy the relationship

$$h = \left[\frac{R(K)}{\mu_2(K)^2 \hat{\psi}_4(\gamma(h)) n} \right]^{1/5},$$

where the pilot bandwidth for the estimation of ψ_4 is a function γ of h. The choice of γ can be motivated by noting the relationship

$$g_{\text{AMSE}} = \left[\frac{2L^{(4)}(0)\mu_2(K)^2}{R(K)\mu_2(L)} \right]^{1/7} (-\psi_4/\psi_6)^{1/7} h_{\text{AMISE}}^{5/7}.$$

This suggests taking

$$\gamma(h) = \left[\frac{2L^{(4)}(0)\mu_2(K)^2}{R(K)\mu_2(L)} \right]^{1/7} \{-\hat{\psi}_4(g_1)/\hat{\psi}_6(g_2)\}^{1/7} h^{5/7}$$

where $\hat{\psi}_4(g_1)$ and $\hat{\psi}_6(g_2)$ are kernel estimates of ψ_4 and ψ_6 (Sheather and Jones, 1991). The choice of g_1 and g_2 can be made using (3.6), although this leads to another "stage selection" problem, as in the direct plug-in case.

EXAMPLE. A two-stage solve-the-equation bandwidth selector that uses $L = K$, which we will denote by $\hat{h}_{\text{STE},2}$, is given below (Sheather and Jones, 1991). It requires a numerical routine to implement Step 4; simulation experience suggests uniqueness of the solution and a relatively easy computational problem to overcome.

Step 1 Estimate ψ_6 and ψ_8 using $\hat{\psi}_6^{\text{NS}} = -15/(16\pi^{1/2}\hat{\sigma}^7)$ and $\hat{\psi}_8^{\text{NS}} = 105/(32\pi^{1/2}\hat{\sigma}^9)$.

Step 2 Estimate ψ_4 and ψ_6 using the kernel estimators $\hat{\psi}_4(g_1)$ and $\hat{\psi}_6(g_2)$ where

$$g_1 = \{-2K^{(4)}(0)/(\mu_2(K)\hat{\psi}_6^{\text{NS}} n)\}^{1/7}$$

and
$$g_2 = \{-2K^{(6)}(0)/(\mu_2(K)\hat{\psi}_8^{NS}n)\}^{1/9}.$$

Step 3 Estimate ψ_4 using the kernel estimator $\hat{\psi}_4(\gamma(h))$ where

$$\gamma(h) = \left[\frac{2K^{(4)}(0)\mu_2(K)\hat{\psi}_4(g_1)}{-\hat{\psi}_6(g_2)R(K)}\right]^{1/7} h^{5/7}.$$

Step 4 The selected bandwidth is the solution to the equation

$$h = \left[\frac{R(K)}{\mu_2(K)^2\hat{\psi}_4(\gamma(h))n}\right]^{1/5}.$$

∎

3.7 Smoothed cross-validation bandwidth selection

Smoothed cross-validation (SCV) (Müller, 1985, Staniswalis, 1989a, Hall, Marron and Park, 1992) is similar to plug-in bandwidth selection in that it uses a kernel estimator with pilot bandwidth g to estimate the integrated squared bias component of MISE$\{\hat{f}(\cdot;h)\}$. Because of this, the methods have similar theoretical properties. The difference is that SCV is based on the exact integrated squared bias rather than its asymptotic approximation. This has the intuitively appealing feature of having less dependence on asymptotic approximations. On the other hand, SCV is not as easy to implement as DPI and is somewhat more difficult to analyse.

In Section 2.3 we showed that

$$\text{MISE}\{\hat{f}(\cdot;h)\} = (nh)^{-1}R(K) + (1 - n^{-1})\int(K_h * f)^2(x)\,dx$$
$$- 2\int(K_h * f)(x)f(x)\,dx + \int f(x)^2\,dx.$$

Ignoring the asymptotically negligible n^{-1} in the second term, we obtain

$$\text{MISE}\{\hat{f}(\cdot;h)\} \simeq (nh)^{-1}R(K) + \int(K_h * f - f)(x)^2\,dx.$$

The second term is exactly the integrated squared bias of $\hat{f}(\cdot;h)$ while the first is a good approximation to the integrated variance. The smoothed cross-validation objective function is obtained by replacing f by a pilot estimator

$$\hat{f}_L(x;g) = n^{-1} \sum_{i=1}^{n} L_g(x - X_i)$$

where $L_g(x) = L(x/g)/g$ for a possibly different kernel L and bandwidth g. This gives us

$$\text{SCV}(h) = (nh)^{-1} R(K) + \widehat{\text{ISB}}(h)$$

where

$$\widehat{\text{ISB}}(h) = \int \{K_h * \hat{f}_L(\cdot;g) - \hat{f}_L(\cdot;g)\}^2(x)\, dx \qquad (3.8)$$

is an estimate of integrated squared bias (ISB). The SCV bandwidth $\hat{h}_{\text{SCV}}$ is defined to be the largest local minimiser of SCV(h). It is a straightforward exercise (Exercise 3.8) to show that $\widehat{\text{ISB}}(h)$ has the more explicit formulation:

$$\widehat{\text{ISB}}(h) = n^{-2} \sum_{i=1}^{n} \sum_{j=1}^{n} (K_h * K_h * L_g * L_g \\ - 2 * K_h * L_g * L_g + L_g * L_g)(X_i - X_j). \qquad (3.9)$$

The SCV bandwidth selector has some interesting alternative motivations (Hall, Marron and Park, 1992). The first is that SCV is an adjustment of least squares cross-validation that allows "pre-smoothing" of the pairwise differences $X_i - X_j$. This is the reason for the method's name and can be seen by noting that if there are no replications among the data then

$$\text{LSCV}(h) = (nh)^{-1} R(K) \\ + n^{-1}(n-1)^{-1} \sum \sum_{i \neq j} (K_h * K_h - 2K_h + K_0)(X_i - X_j)$$

where K_0 is the Dirac delta function. A variation of SCV(h) is one where the integrated squared bias is instead estimated by

$$\widetilde{\text{ISB}}(h) = n^{-1} \sum_{i=1}^{n} \int \{K_h * \hat{f}_{L,-i}(x;g) - \hat{f}_{L,-i}(x;g)\}^2$$

3.7 SMOOTHED CROSS-VALIDATION BANDWIDTH SELECTION

where $\hat{f}_{L,-i}(\cdot;g)$ is the leave-one-out kernel density estimator based on kernel L and bandwidth g. Then

$$\widetilde{\text{ISB}}(h) = \{n(n-1)\}^{-1} \sum\sum_{i\neq j} [\{K_h * K_h - 2K_h + K_0\} * L_g * L_g](X_i - X_j).$$

Thus, this version of SCV estimates ISB in the same way as LSCV except that the function $L_g * L_g$ is applied to the $X_i - X_j$. This also makes it clear that LSCV is a special case of the leave-one-out version of SCV with $g = 0$.

SCV can also be motivated through the smoothed bootstrap. Let $X_1^*, \ldots, X_n^*$ be a sample from the density $\hat{f}_L(\cdot;g)$ conditional on the original sample $X_1, \ldots, X_n$. The bootstrap estimate of MISE$\{\hat{f}(\cdot;h)\}$ is then

$$\text{MISE}^*\{\hat{f}^*(\cdot;h)\} = E^* \int \{\hat{f}^*(x;h) - \hat{f}_L(x;g)\}^2 \, dx$$

where

$$\hat{f}^*(x;h) = n^{-1} \sum_{i=1}^n K_h(x - X_i^*)$$

and E^* denotes expectation conditional on $X_1, \ldots, X_n$. However, one can show (Exercise 3.8) that

$$\text{MISE}^*\{\hat{f}^*(\cdot;h)\} = \text{SCV}(h)$$
$$+ n^{-1} \int \{K_h * \hat{f}_L(\cdot;g)(x)\}^2 \, dx. \qquad (3.10)$$

The second term is asymptotically negligible with respect to the first, so choosing h to minimise $\text{MISE}^*\{\hat{f}^*(\cdot;h)\}$ is virtually equivalent to choosing the bandwidth that minimises $\text{SCV}(h)$.

Alternative approaches to SCV, based on the frequency domain version of MISE, have also been proposed and analysed (Chiu, 1991, 1992).

Considerable theory has been devoted to the choice of the pilot bandwidth g in the SCV objective function (Jones, Marron and Park, 1991, Park and Marron, 1992), and this involves precisely the same considerations as were necessary for the selection of a pilot bandwidth of a plug-in bandwidth selector. There are good asymptotic reasons for allowing g to have dependence on h of the form

$$g = Cn^p h^m$$

for constants C, p and m. Optimal choice of these constants, aimed at enhancing the asymptotic performance of $\hat{h}_{\text{SCV}}$, will be discussed in Section 3.8. (In parallel with this, direct plug-in rules take $(m,p) = (0, \frac{1}{7})$ and solve-the-equation rules take $(m,p) = (\frac{5}{7}, 0)$.) Asymptotically optimal choice of C depends on f through ψ_r functionals, so these can be estimated in the same way as was done for plug-in bandwidth selection algorithms. A smoothed cross-validation bandwidth selector that involves ℓ successive functional estimation steps, with an initial bandwidth chosen by a quick and simple rule, will be denoted by $\hat{h}_{\text{SCV},\ell}$.

The following example describes one particular version of $\hat{h}_{\text{SCV},2}$.

EXAMPLE. Let $K = L = \phi$, the normal kernel.

Step 1 Compute kernel estimates $\hat{\psi}_6(g_1)$ and $\hat{\psi}_{10}(g_2)$ where

$$g_1 = \{2/(7n)\}^{1/9} 2^{1/2} \hat{\sigma} \quad \text{and} \quad g_2 = \{2/(11n)\}^{1/13} 2^{1/2} \hat{\sigma}$$

are based on normal scale estimates of ψ_8 and ψ_{12}; see (3.7).

Step 2 Compute kernel estimates $\hat{\psi}_4(g_3)$ and $\hat{\psi}_8(g_4)$ where

$$g_3 = [-6/\{(2\pi)^{1/2} \hat{\psi}_6(g_1) n\}]^{1/7}$$

and

$$g_4 = [-210/\{(2\pi)^{1/2} \hat{\psi}_{10}(g_2) n\}]^{1/11}.$$

Step 3 Choose h to minimise

$$\text{SCV}(h) = (nh)^{-1}(2\pi^{1/2})^{-1} + \sum_{i=1}^{n}\sum_{j=1}^{n}\{\phi_{(2h^2+2g^2)^{1/2}} - 2\phi_{(h^2+2g^2)^{1/2}} + \phi_{(2g^2)^{1/2}}\}(X_i - X_j)$$

where

$$g = \hat{C} n^{-23/45} h^{-2}$$

and

$$\hat{C} = \{441/(64\pi)\}^{1/18} (4\pi)^{-1/5} \hat{\psi}_4(g_3)^{-2/5} \hat{\psi}_8(g_4)^{-1/9}.$$

3.8 COMPARISON OF BANDWIDTH SELECTORS

In the above example we took both K and L to be the normal kernel to keep the presentation simple. Note that the identity

$$(\phi_\sigma * \phi_{\sigma'})(x) = \phi_{(\sigma^2+\sigma'^2)^{1/2}}(x)$$

(Appendix C) has been used to simplify the convolution. An explanation for the particular choices of $\hat{C}$, p and m used in this example is given in the next section.

3.8 Comparison of bandwidth selectors

As the previous four sections indicate, there now exist several hi-tech rules for selecting the bandwidth of a kernel density estimator from the data. This leads to the natural question: how do these methods compare? Some insight into this can be obtained through asymptotic analysis of each bandwidth selector which typically leads to a "rate of convergence" of the selected bandwidth to some optimal bandwidth. While such a theoretical result gives some indication of the relative merits of competing bandwidth selectors, it does not necessarily indicate what happens in practice since the asymptotics often do not take effect for smaller sample sizes. Therefore, computer simulation has also become an important tool for the comparison of bandwidth selectors.

3.8.1 Theoretical performance

The usual way of quantifying the theoretical performance of a particular bandwidth selector $\hat{h}$ is through an asymptotic distribution result, typically of the form

$$n^\nu(\hat{h}/h_0 - 1) \to_D N(\mu, \sigma^2) \qquad (3.11)$$

where μ and $\sigma^2 > 0$ depend only on f and K (but not on n) and h_0 is some "optimal" bandwidth. The notation

$$A_n \to_D N(\mu, \sigma^2)$$

means that the sequence of random variables A_n converges in distribution to a $N(\mu, \sigma^2)$ random variable (see Appendix A). The quantity $\hat{h}/h_0 - 1$ is called the *relative error* of $\hat{h}$ and takes into account the fact that the bandwidth is a scale parameter. If $\hat{h}$ satisfies (3.11) then we will say that $\hat{h}$ has a *relative rate of convergence* to h_0 of order $n^{-\nu}$ with asymptotic variance σ^2. Therefore, larger values of ν correspond to the bandwidth selectors converging to the optimum at faster rates. In addition, smaller values of σ^2 correspond to the bandwidth selectors being more stable.

An important question is that of the most appropriate choice for the "optimal" bandwidth h_0. Since each of the bandwidths in this section is based on minimising an estimate of $\text{MISE}\{\hat{f}(\cdot;h)\}$, a natural first answer might be $h_0 = h_{\text{MISE}}$. Observe, however, that h_{MISE} is the bandwidth minimising a quantity that is *averaged over all possible samples* while the optimal bandwidth for the *sample at hand* (with respect to squared error loss) is h_{ISE}, the bandwidth that minimises

$$\text{ISE}\{\hat{f}(\cdot;h)\} = \int \{\hat{f}(x;h) - f(x)\}^2\, dx.$$

In asymptotic distribution terms, the choices $h_0 = h_{\text{MISE}}$ and $h_0 = h_{\text{ISE}}$ give quite different results. One notable manifestation of this is through minimax results that state that the relative rate of convergence to h_{ISE} of any bandwidth selector cannot be faster than $n^{-1/10}$ while the relative rate of convergence of bandwidth selectors to h_{MISE} can be as fast as $n^{-1/2}$ (Hall and Marron, 1991b). These results indicate that the goal of minimising $\text{ISE}\{\hat{f}(\cdot;h)\}$ is much more difficult than that of minimising $\text{MISE}\{\hat{f}(\cdot;h)\}$, due to the fact that h_{ISE} is, itself, a random variable. For essentially this reason (see also Jones, 1991a, and Grund, Hall and Marron, 1995) we will take $h_0 = h_{\text{MISE}}$ for our theoretical comparison.

Least squares cross-validation and biased cross-validation

Under certain regularity conditions the LSCV bandwidth selector satisfies

$$n^{1/10}(\hat{h}_{\text{LSCV}}/h_{\text{MISE}} - 1) \to_D N(0, \sigma^2_{\text{LSCV}})$$

(Hall and Marron, 1987a; Scott and Terrell, 1987) while the BCV bandwidth selector satisfies

$$n^{1/10}(\hat{h}_{\text{BCV}}/h_{\text{MISE}} - 1) \to_D N(0, \sigma^2_{\text{BCV}})$$

(Scott and Terrell, 1987). Here σ^2_{LSCV} and σ^2_{BCV} depend on functionals of f and K, but not on n. The ratio of these two asymptotic variances for the standard normal kernel is

$$\sigma^2_{\text{LSCV}}/\sigma^2_{\text{BCV}} \simeq 15.7.$$

3.8 COMPARISON OF BANDWIDTH SELECTORS

This indicates that we should expect $\hat{h}_{\text{LSCV}}$ to be considerably more variable than $\hat{h}_{\text{BCV}}$, which is a theoretical explanation of the phenomenon observed in Figure 3.3.

The above results also show that both $\hat{h}_{\text{LSCV}}$ and $\hat{h}_{\text{BCV}}$ have a relative rate of convergence to h_{MISE} of order $n^{-1/10}$, which is considerably slower than the $n^{-1/2}$ lower bound. An understanding of what causes this poor asymptotic performance can be most easily gained by a comparison of BCV and a direct plug-in bandwidth selector. Recall that $\hat{h}_{\text{BCV}}$ minimises

$$\text{BCV}(h) = (nh)^{-1}R(K) + \tfrac{1}{4}h^4\mu_2(K)^2\widetilde{\psi}_4(h),$$

while a slight variant of the direct plug-in bandwidth selector, which we will call $\tilde{h}_{\text{DPI}}$, minimises

$$\text{DPI}(h) = (nh)^{-1}R(K) + \tfrac{1}{4}h^4\mu_2(K)^2\widetilde{\psi}_4(g),$$

where g is independent of h. Here

$$\widetilde{\psi}_4(g) = n^{-1}(n-1)^{-1}\sum\sum_{i \neq j}(K_g'' * K_g'')(X_i - X_j)$$

is a "leave-out-diagonals" estimate of ψ_4. The two methods differ in their choice of the pilot bandwidth g. The selector $\hat{h}_{\text{BCV}}$ overcomes this choice by setting $g = h$ while $\tilde{h}_{\text{DPI}}$ allows for arbitrary choice of g. Theory similar to that presented in Section 3.5 shows that optimal choice of g, with respect to asymptotic MSE$\{\widetilde{\psi}_4(g)\}$, is of order $n^{-2/13}$ whereas the h used by BCV is asymptotically of order $n^{-1/5}$. This suboptimal choice of h for estimation of the integrated squared bias approximation is the reason for the asymptotic deficiency of BCV. Similar arguments can be used to explain the poor asymptotic performance of LSCV.

Plug-in and smoothed cross-validation selectors

As the preceding discussion suggests, the extra flexibility due to having a pilot bandwidth for estimation of integrated squared bias allows for considerable improvement of the asymptotic relative error of plug-in and smoothed cross-validation bandwidth selectors. We will now study the effect of g on the relative error of $\hat{h}_{\text{DPI}}$. This will be followed by similar results for $\hat{h}_{\text{SCV}}$.

The direct plug-in bandwidth selector can be written

$$\hat{h}_{\text{DPI}} = \alpha(K)\hat{\psi}_4(g)^{-1/5}n^{-1/5}$$

where $\alpha(K) = \{R(K)/\mu_2(K)^2\}^{1/5}$. An analysis of the errors involved in the approximation of h_{MISE} by h_{AMISE} leads to

$$h_{\text{MISE}} = \alpha(K)\psi_4^{-1/5}n^{-1/5} + O(n^{-3/5}).$$

The relative error of $\hat{h}_{\text{DPI}}$ is therefore

$$\hat{h}_{\text{DPI}}/h_{\text{MISE}} - 1 = \psi_4^{-1/5}\{\hat{\psi}_4(g)^{-1/5} - \psi_4^{-1/5}\} + O_P(n^{-2/5})$$

where O_P denotes *order in probability* (see Appendix A) and takes into account the randomness of $\hat{\psi}_4(g)$. The $O_P(n^{-2/5})$ term is due to the relative error of h_{AMISE} as an approximation to h_{MISE}. Formal application of Taylor's theorem leads to

$$\hat{h}_{\text{DPI}}/h_{\text{MISE}} - 1 = -\tfrac{1}{5}\psi_4^{-1}\{\hat{\psi}_4(g) - \psi_4\} + \ldots + O_P(n^{-2/5}).$$

This expression shows that the relative error of $\hat{h}_{\text{DPI}}$ depends on the functional estimation error $\hat{\psi}_4(g) - \psi_4$ as well as the relative approximation error of h_{AMISE}. Either term can dominate, depending on the choice of L and g. For the bandwidth selector $\hat{h}_{\text{DPI},2}$ given in the example of Section 3.6.1, which is based on $L = K$ being a second-order kernel with g chosen to optimise $\text{AMSE}\{\hat{\psi}_4(g)\}$, it can be shown that $\hat{\psi}_4(g) - \psi_4$ is $O_P(n^{-5/14})$ which dominates the $O_P(n^{-2/5})$ term. More involved and rigorous arguments can be used to show that we in fact have

$$n^{5/14}(\hat{h}_{\text{DPI},2}/h_{\text{MISE}} - 1) \to_D N(0, \sigma_{\text{DPI}}^2)$$

so this particular version of $\hat{h}_{\text{DPI}}$, with a relative convergence rate of order $n^{-5/14}$, is considerably closer to the $n^{-1/2}$ lower bound than the $n^{-1/10}$ rate of cross-validation methods. The same result can also be shown to hold for $\hat{h}_{\text{STE},2}$.

The role of the pilot bandwidth g for $\hat{h}_{\text{SCV}}$ is analogous to that for $\hat{h}_{\text{DPI}}$, although theory for optimal choice of g is not as simple to derive. Asymptotic distributional results can be used to provide insight into appropriate choice of g. Let $\hat{h}_{\text{SCV}}$ be the minimiser of $\text{SCV}(h)$, where $g = Cn^p h^m$. Also, let $g_{\text{MISE}} = Cn^p h_{\text{MISE}}^m$. If K and L are both second-order kernels then, under appropriate regularity conditions, we have

$$\begin{aligned}\hat{h}_{\text{SCV}}/h_{\text{MISE}} - 1 =\ & (n^{-4/5}h_{\text{MISE}}^6 g_{\text{MISE}}^{-9} C_1 + n^{-1}C_2)^{1/2}Z \\ & + (-n^{3/5}h_{\text{MISE}}^3 g_{\text{MISE}}^2 C_3 + n^{3/5}h_{\text{MISE}}^3 g_{\text{MISE}}^4 C_4 \\ & + n^{-2/5}h_{\text{MISE}}^3 g_{\text{MISE}}^{-5} C_5)\end{aligned} \quad (3.12)$$

(Jones, Marron and Park, 1991), where Z is asymptotically $N(0,1)$. An expression of this type also exists for $\hat{h}_{\text{DPI}}$ and $\hat{h}_{\text{STE}}$, which confirms that the choice of pilot bandwidth is analogous. The constants $C_1, \ldots, C_5$ depend on functionals of K and L, and on f only through ψ_r density functionals, an exception being C_2 which also depends on $\int \{f^{(4)}(x)\}^2 f(x)\,dx$. Of particular relevance is the expression for C_3:

$$C_3 = \tfrac{1}{10}(m+2)\left[\frac{-\mu_2(K)^6 \mu_2(L)^5 \psi_6^5}{R(K)^3 \psi_4^2}\right]^{1/5}.$$

This is because the choice $m = -2$ leads to $C_3 = 0$. The first term on the right hand side of (3.12) can be thought of as representing the variation of $\hat{h}_{\text{SCV}}$, while the second represents bias. This shows how choice of g represents a type of variance-bias trade-off. Bias is decreased by having $g \to 0$, but this decrease should not be too rapid, otherwise the variance is increased. A simple means of determining the optimal choice of g is to combine the asymptotic variance and bias terms to obtain an "asymptotic relative mean squared error (ARMSE)",

$$\begin{aligned}\text{ARMSE} = {}& n^{-4/5} h_{\text{MISE}}^6 g_{\text{MISE}}^{-9} C_1 + n^{-1} C_2 \\ & + (-n^{3/5} h_{\text{MISE}}^3 g_{\text{MISE}}^2 C_3 + n^{3/5} h_{\text{MISE}}^3 g_{\text{MISE}}^4 C_4 \\ & + n^{-2/5} h_{\text{MISE}}^3 g_{\text{MISE}}^{-5} C_5)^2.\end{aligned}$$

If $m \neq -2$ then the term involving C_3 dominates that involving C_4. Choosing g to cancel this term asymptotically with the term involving C_5 leads to

$$g = \{(C_5/C_3)^{1/7}/C_0^m\} n^{m/5 - 1/7} h^m$$

being the asymptotically optimal choice, where

$$C_0 = [R(K)/\{\mu_2(K)^2 R(f'')\}]^{1/5}$$

and arises from the fact that $h_{\text{MISE}} \sim C_0 n^{-1/5}$. If $m = 0$ (so g is independent of h) we simply have

$$g = (C_5/C_3)^{1/7} n^{-1/7}$$

which is essentially the same as the optimal pilot bandwidth g_2 for the direct plug-in bandwidth selector described in the example of

Section 3.6.1. The resulting rate of convergence of $\hat{h}_{\text{SCV}}$ is given by

$$n^{5/14}(\hat{h}_{\text{SCV}}/h_{\text{MISE}} - 1) \to_D N(0, \sigma^2_{\text{SCV}}).$$

Root-n bandwidth selection

It is possible to construct plug-in and smooth cross-validation rules that achieve the optimal $n^{-1/2}$ relative rate of convergence. Such rules are called *root-n bandwidth selectors*.

One of the simplest root-n bandwidth selectors is based on the following two term approximation to h_{MISE},

$$h_{\text{AMISE},2} = \left[\frac{R(K)}{\mu_2(K)^2 \psi_4 n}\right]^{1/5} + \frac{\mu_4(K)\psi_6}{20}\left[\frac{R(K)^3}{\mu_2(K)^{11}\psi_4^8 n^3}\right]^{1/5}.$$

The relative error of $h_{\text{AMISE},2}$ is of order $n^{-3/5}$, compared to the $n^{-2/5}$ relative rate of the usual one term approximation, h_{AMISE}. Thus, if higher-order kernels are used to estimate ψ_4 and ψ_6 with mean squared error of $O(n^{-1})$ then the resulting plug-in bandwidth selector achieves root-n performance (Hall, Sheather, Jones and Marron, 1991).

The root-n relative rate can also be achieved by judicious choice of the parameters via the $g = Cn^p h^m$ factorization of the pilot bandwidth. We will describe this in the SCV context, although similar results are obtainable for plug-in bandwidth selection. If $m = -2$ then C_3 as defined above vanishes and the bias term is asymptotically zeroed by

$$g = (-C_5/C_4)^{1/9} C_0^2 n^{-23/45} h^{-2}.$$

This leads to

$$n^{1/2}(\hat{h}_{\text{SCV}}/h_{\text{MISE}} - 1) \to_D N(0, \sigma^2_{\text{SCV}})$$

(with a different value for σ^2_{SCV} than above) (Jones, Marron and Park, 1991). An appealing feature of this result is that the $n^{-1/2}$ rate can be obtained without the use of higher-order kernels at any stage. The example of Section 3.7 describes the rule given by this strategy when the normal kernel is used at all stages.

Root-n performance has also been established for the frequency domain approach to SCV (Chiu, 1991, 1992).

3.8.2 Practical advice

The asymptotic results of the previous section need to be viewed with some caution. Apart from requiring that the sample size be sufficiently large they also have the defect of often masking the choice of various auxiliary parameters, such as the choice of scale estimate for a normal scale rule, or the number of stages of a plug-in strategy. These parameters can have a significant effect on the performance of a bandwidth selector in practice.

The main tool for assessing the practical performance of a bandwidth selector is simulation. Figure 3.4, for example, shows that important insight into the effect of the number of stages of a plug-in rule can be obtained from simulation. Figure 3.5 provides a similar comparison of the selectors $\hat{h}_{\text{LSCV}}$, $\hat{h}_{\text{BCV}}$ (with K equal to the standard normal kernel) and the versions of $\hat{h}_{\text{DPI},2}$ and $\hat{h}_{\text{SCV},2}$ given in the examples of Section 3.6.1 and Section 3.7 respectively. The sample size is $n = 100$ and the underlying density is f_1 as defined at (2.3).

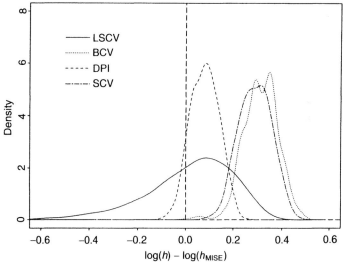

Figure 3.5. *Density estimates based on values of* $\log_{10}(\hat{h}) - \log_{10}(h_{\text{MISE}})$ *for several bandwidth selectors (described in the text). Selected bandwidths are based on 500 simulated samples of size* $n = 100$ *from the normal mixture density* f_1 *defined at (2.3).*

For this particular setting we see that $\hat{h}_{\text{DPI},2}$ provides the best compromise between bias and variance among these four selectors.

It is difficult to give a concise summary of simulation results in general since the rankings of the selectors change for different densities (see e.g. Park and Turlach (1992), Cao, Cuevas and González-Manteiga (1994), Jones, Marron and Sheather (1992)), although certain patterns have emerged. As the asymptotics suggest, versions of hi-tech selectors involving pilot bandwidths, such as $\hat{h}_{\mathrm{DPI}}$ and $\hat{h}_{\mathrm{SCV}}$, perform quite well for many density types (e.g. Park and Turlach, 1992). However, for densities with sharp features, such as several modes, the asymptotics that these methods rely upon tend to be less accurate and the performance can be worse than a non-asymptotic method. LSCV does not depend on asymptotic arguments, but its sample variability is usually considered to be too high to be of reliable practical use (e.g. Jones, Marron and Sheather, 1992). The simulation performance of BCV has also been disappointing and we cannot recommend this bandwidth selector for general use.

In summary, while considerable recent progress has been made in the development towards high-performance bandwidth selectors, no rule comes with a guarantee that it will work satisfactorily in all cases. A sensible data analytic strategy is that of obtaining estimates for a variety of bandwidths, perhaps obtained from a variety of bandwidth selectors and choices of auxiliary parameters. If a single objective bandwidth selector is required then our recommendation, based on simulation evidence, is to use a version of $\hat{h}_{\mathrm{DPI}}$, $\hat{h}_{\mathrm{STE}}$ or $\hat{h}_{\mathrm{SCV}}$, rather than $\hat{h}_{\mathrm{LSCV}}$ or $\hat{h}_{\mathrm{BCV}}$.

3.9 Bibliographical notes

3.2 For a discussion of normal scale rules based on simple scale measures see Silverman (1986, pp.45–47). Bowman (1985) showed their usefulness for normal-like densities. Janssen, Marron, Veraverbeke and Sarle (1995) developed improved scale measures for use in bandwidth selection. Oversmoothed bandwidth selection methods were proposed and discussed in Terrell and Scott (1985) and Terrell (1990). Sheather (1992) applied several bandwidth selectors to the Old Faithful data.

3.3 Rudemo (1982) and Bowman (1984) independently derived LSCV. Notable contributions to the theory of LSCV include Hall (1983), Stone (1984) and Hall and Marron (1987a). Related cross-validatory smoothing parameter selectors for Fourier series density estimates were proposed by Kronmal and Tarter (1968) and Wahba (1981).

3.4 The BCV bandwidth selector was proposed by Scott and

3.9 BIBLIOGRAPHICAL NOTES

Terrell (1987). A theoretical comparison of LSCV, BCV and related bandwidth selectors was given by Jones and Kappenman (1992).

3.5 The essence of the theory given here was first developed by Hall and Marron (1987b). A modification of their estimator, with improved performance, was proposed by Jones and Sheather (1991) and is studied in this section. Aldershof (1991) contains a very detailed study of estimators of this type.

3.6 The idea of plug-in bandwidth selection dates back to Woodroofe (1970) and Nadaraya (1974), although these were theoretical contributions that did not address the practical choice of the pilot bandwidth. More practical rules of the solve-the-equation variety were proposed by Scott, Tapia and Thompson (1977) (see also Scott and Factor, 1981) and Sheather (1983, 1986). Refinements have been proposed by Park and Marron (1990), Sheather and Jones (1991), Hall and Johnstone (1992) and Engel, Herrmann and Gasser (1995).

3.7 Smoothed cross-validation as described in this section was developed by Hall, Marron and Park (1992). However, similar ideas were employed in the kernel regression context by Müller (1985) and Staniswalis (1989a). Bootstrap bandwidth selectors, equivalent to versions of SCV, were proposed by Taylor (1989) and Faraway and Jhun (1990). Contributions to the theoretical properties of SCV include Jones, Marron and Park (1991), Park and Marron (1992), Chiu (1991, 1992) and Kim, Park and Marron (1994).

3.8 Asymptotic distributions of the various bandwidth selectors considered here were first derived by Hall and Marron (1987a), Scott and Terrell (1987), Park and Marron (1990), Sheather and Jones (1991), Jones, Marron and Park (1991), Hall, Marron and Park (1992) and Park and Marron (1992). The issue concerning whether h_{ISE} or h_{MISE} is more appropriate for the optimal bandwidth has been discussed by Mammen (1990), Jones (1991a) and Grund, Hall and Marron (1995). Root-n bandwidth selectors have been proposed by Hall, Sheather, Jones and Marron (1991), Jones, Marron and Park (1991), Chiu (1991, 1992) and Kim, Park and Marron (1994). Marron (1989) provides some useful results for measuring Monte Carlo error in bandwidth selection simulation studies. Recent comparisons of the practical performance of bandwidth selectors, through either simulation studies or real data examples, can be found in Park and Marron (1990), Chiu (1992), Park and Turlach (1992), Sheather (1992), Jones, Marron and Sheather

(1992) and Cao, Cuevas and González-Manteiga (1994). The last two references are also reviews of the subject.

3.10 Exercises

3.1 Using (3.7) show that the AMISE-optimal bandwidth for estimating the $N(0,\sigma^2)$ density is given by (3.1).

3.2 Let
$$f(x) = \tfrac{35}{32}(1-x^2)^3 1_{\{|x|<1\}}$$
be the triweight density, and
$$f_\mu(x) = \tfrac{1}{2}f(x+\mu) + \tfrac{1}{2}f(x-\mu), \quad \mu > 1$$
be a family of bimodal densities based on f. For density f_μ compute $\hat{h}_{\text{AMISE}}$ and $\hat{h}_{\text{NS}}$ with
(a) $\hat{\sigma}$ equal to the population standard deviation,
(b) $\hat{\sigma}$ equal to the population version of $\hat{\sigma}_{\text{IQR}}$.
Assuming that the sample standard deviation and $\hat{\sigma}_{\text{IQR}}$ provide good estimates of their population counterparts argue that $\hat{h}_{\text{NS}}$ based on either s or $\hat{\sigma}_{\text{IQR}}$ can be made to perform arbitrarily poorly by taking μ to be large enough.

3.3 Show that
$$E\{\text{LSCV}(h)\} = \text{MISE}\{\hat{f}(\cdot;h)\} - \int f(x)^2 \, dx.$$

3.4 Let $X_1, \ldots, X_n$ be a set of identically and independently distributed random variables and define
$$U = 2n^{-2} \sum_{i=1}^{n-1} \sum_{j=i+1}^{n} S(X_i - X_j)$$
where the function S is symmetric about zero. Show that
$$\text{Var}(U) = 2n^{-3}(n-1)\text{Var}\{S(X_1 - X_2)\}$$
$$+ 4n^{-3}(n-1)(n-2)\text{Cov}\{S(X_1 - X_2), S(X_2 - X_3)\}.$$

3.5 Let f be the standard normal density and consider the problem of estimation of $\psi_4 = 3/(8\pi^{1/2})$ using the kernel estimator
$$\hat{\psi}_4(g) = n^{-2} \sum_{i=1}^{n} \sum_{j=1}^{n} L_g^{(4)}(X_i - X_j).$$

3.10 EXERCISES

Show that the AMSE-optimal choice of g when L is the standard normal kernel is

$$g_{\text{AMSE}} = \left[\frac{16(2^{1/2})}{5n}\right]^{1/7}$$

and the corresponding optimal $\text{AMSE}\{\hat{\psi}_4(g)\}$ is equal to

$$\frac{105}{32\pi}\left[\frac{5(2^{1/2})}{32}\right]^{9/7} n^{-5/7}.$$

You may use (3.7) and results from Appendix C.

3.6 Suppose ψ_r were estimated by

$$\tilde{\psi}_r(g) = \hat{\psi}_r(g) - n^{-1}L_g^{(r)}(0) = n^{-2}\sum\sum_{i\neq j}L_g^{(r)}(X_i - X_j),$$

instead of $\hat{\psi}_r(g)$. Determine the resulting $\text{AMSE}\{\tilde{\psi}_r(g)\}$, the optimal value of g, and corresponding optimal rate of convergence of $\text{MSE}\{\tilde{\psi}_r(g)\}$.

3.7 An important quantity in nonparametric statistics is $\psi_0 = \int f(x)^2\,dx$. Use the theory of Section 3.5 to determine the asymptotic MSE properties of the kernel estimator for ψ_0 based on a second-order kernel L. In particular, derive the AMSE-optimal bandwidth and corresponding MSE optimal rate of convergence of this kernel estimator of ψ_0 (Sheather, Hettmansperger and Donald, 1994).

3.8
(a) Show that the expression for $\widehat{\text{ISB}}(h)$ given by (3.8) can be written as (3.9).
(b) Show that the bootstrap estimate of $\text{MISE}\{\hat{f}(\cdot;h)\}$ equals the right-hand side of (3.10).

3.9 Write out the set of steps required for $\hat{h}_{\text{DPI},1}$ based on a normal scale rule and with $K = L$, a second-order kernel.

3.10 The AMISE-optimal binwidth for the histogram is

$$b_{\text{AMISE}} = \{6/(-\psi_2)\}^{1/3} n^{-1/3}$$

(Scott, 1979). Use this result and (3.7) to derive
(a) a quick and simple rule for selecting b,
(b) a one-stage direct plug-in binwidth selector for b where a kernel estimator is used to estimate ψ_2.

CHAPTER 4

Multivariate kernel density estimation

4.1 Introduction

We will now investigate the extension of the kernel density estimator to the multivariate setting. The need for nonparametric density estimates for recovering structure in multivariate data is, perhaps, greater since parametric modelling is more difficult than in the univariate case. However, the extension of the univariate kernel methodology discussed in Chapters 1 and 2 is not without its problems. The most general smoothing parametrisation of the kernel estimator in higher dimensions requires the specification of many more bandwidth parameters than in the univariate setting. This leads us to consider simpler smoothing parametrisations as well. Also, the sparseness of data in higher-dimensional space makes kernel smoothing difficult unless the sample size is very large. This phenomenon, usually called the *curse of dimensionality*, means that, with practical sample sizes, reasonable nonparametric density estimation is very difficult in more than about five dimensions (see Exercise 4.1). Nevertheless, there have been several studies where the kernel density estimator has been an effective tool for displaying structure in bivariate samples (e.g. Silverman, 1986, Scott, 1992). The multivariate kernel density estimate has also played an important role in recent developments of the visualisation of structure in three- and four-dimensional data sets (Scott, 1992).

The multivariate kernel density estimator that we study in this chapter is a direct extension of the univariate estimator. We should note that there are also approaches to multivariate smoothing which attempt to alleviate the curse of dimensionality by assuming that the multivariate function has some simplifying structure (e.g. Friedman, Stuetzle and Schroeder, 1984).

Throughout this chapter we will let $\mathbf{X}_1, \ldots, \mathbf{X}_n$ denote a d-variate random sample having density f. We will use the notation $\mathbf{X}_i = (X_{i1}, \ldots, X_{id})^T$ to denote the components of $\mathbf{X}_i$ and a generic vector $\mathbf{x} \in \mathbb{R}^d$ will have the representation $\mathbf{x} = (x_1, \ldots, x_d)^T$. Also, $\int$ will be shorthand for $\int \cdots \int_{\mathbb{R}^d}$ and $d\mathbf{x}$ will be shorthand for $dx_1 \cdots dx_d$. Finally, the $d \times d$ identity matrix will be denoted by $\mathbf{I}$.

4.2 The multivariate kernel density estimator

In its most general form, the d-dimensional kernel density estimator is

$$\hat{f}(\mathbf{x}; \mathbf{H}) = n^{-1} \sum_{i=1}^{n} K_{\mathbf{H}}(\mathbf{x} - \mathbf{X}_i) \tag{4.1}$$

(Deheuvels, 1977b) where $\mathbf{H}$ is a symmetric positive definite $d \times d$ matrix called the *bandwidth matrix*,

$$K_{\mathbf{H}}(\mathbf{x}) = |\mathbf{H}|^{-1/2} K(\mathbf{H}^{-1/2} \mathbf{x})$$

and K is a d-variate kernel function satisfying

$$\int K(\mathbf{x}) \, d\mathbf{x} = 1.$$

The kernel function is often taken to be a d-variate probability density function. There are two common techniques for generating multivariate kernels from a symmetric univariate kernel κ:

$$K^P(\mathbf{x}) = \prod_{i=1}^{d} \kappa(x_i) \quad \text{and} \quad K^S(\mathbf{x}) = c_{\kappa,d} \kappa\{(\mathbf{x}^T \mathbf{x})^{1/2}\}$$

where $c_{\kappa,d}^{-1} = \int \kappa\{(\mathbf{x}^T \mathbf{x})^{1/2}\} \, d\mathbf{x}$. The first of these is often called a *product* kernel, the second has the property of being *spherically* or *radially symmetric*. A popular choice for K is the standard d-variate normal density

$$K(\mathbf{x}) = (2\pi)^{-d/2} \exp(-\tfrac{1}{2} \mathbf{x}^T \mathbf{x})$$

in which case $K_{\mathbf{H}}(\mathbf{x} - \mathbf{X}_i)$ is the $N(\mathbf{X}_i, \mathbf{H})$ density in the vector $\mathbf{x}$. The normal kernel can be constructed from the univariate standard normal density using either the product or spherically symmetric extensions.

Let $\mathcal{F}$ denote the class of symmetric, positive definite $d \times d$ matrices. In general $\mathbf{H}$ has $\frac{1}{2}d(d+1)$ independent entries which, even for moderate d, can be a substantial number of smoothing parameters to have to choose. A simplification of (4.1) can be obtained by imposing the restriction $\mathbf{H} \in \mathcal{D}$, where $\mathcal{D} \subseteq \mathcal{F}$ is the subclass of diagonal positive definite $d \times d$ matrices. Then for $\mathbf{H} \in \mathcal{D}$, $\mathbf{H} = \text{diag}(h_1^2, \ldots, h_d^2)$. For $\mathbf{H} \in \mathcal{D}$ the kernel estimator can then be written

$$\hat{f}(x; \mathbf{h}) = n^{-1} \left(\prod_{\ell=1}^{d} h_\ell \right)^{-1} \sum_{i=1}^{n} K \left(\frac{x_1 - X_{i1}}{h_1}, \ldots, \frac{x_d - X_{id}}{h_d} \right)$$

(Epanechnikov, 1969). A further simplification follows from the restriction $\mathbf{H} \in \mathcal{S}$, where $\mathcal{S} = \{h^2 \mathbf{I} : h > 0\}$ and leads to the single bandwidth kernel estimator

$$\hat{f}(\mathbf{x}; h) = n^{-1} h^{-d} \sum_{i=1}^{n} K\{(\mathbf{x} - \mathbf{X}_i)/h\} \qquad (4.2)$$

(Cacoullos, 1966). Thus, we see that there is a hierarchical class of smoothing parametrisations from which to choose when using a multivariate kernel estimator. We will discuss the implications of this choice in detail in Section 4.6.

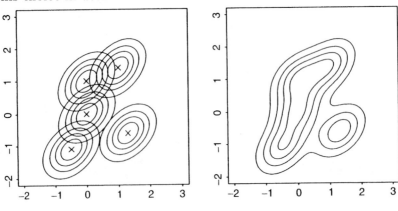

(a) (b)

Figure 4.1. *Construction of a bivariate kernel density estimate: (a) kernel mass being centred about each observation, (b) contour diagram of the resulting kernel estimate.*

4.2 THE MULTIVARIATE KERNEL DENSITY ESTIMATOR

Figure 4.1 shows how a bivariate kernel density estimate is constructed. In Figure 4.1 (a) the crosses represent values of a bivariate random sample. The kernel estimator is then formed by centring a bivariate kernel function around each point. These kernels are represented by the elliptical contours. The heights of these contours are then averaged to form the kernel density estimate which is shown in contour form in Figure 4.1 (b).

Figure 4.2 (a) is a scatterplot of a bivariate data set consisting of 640 longitude/latitude pairs of the epicentres of the earthquakes in the Mount Saint Helens area of the United States (source: O'Sullivan and Pawitan, 1993). The kernel density estimate of these data using the Gaussian kernel and the bandwidth matrix

$$\mathbf{H} = \begin{bmatrix} 0.005^2 & 0 \\ 0 & 0.004^2 \end{bmatrix}$$

is shown in Figure 4.2 (b).

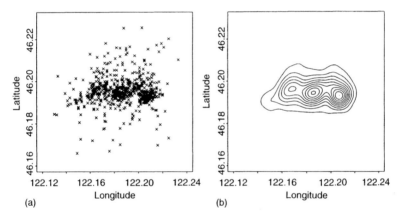

Figure 4.2. *(a) Scatterplot and (b) contour diagram of a kernel density estimate of the Mount Saint Helens earthquake data.*

The three modes indicate the major centres of earthquake activity. Note that the modes are not as apparent from the scatterplot which exemplifies the usefulness of bivariate density estimates for highlighting structure. Scatterplots are the most widely used means of graphically displaying bivariate data sets. However, they have the disadvantage that the eye is drawn to the peripheries of the data cloud, while structure in the main body of the data will tend to be obscured by the high density of points. Kernel density estimates such as that in Figure 4.2 (b) have a clear advantage in presenting information about the joint distribution of the data.

Just as in the univariate case, some important choices have to be made when constructing a multivariate kernel density estimator. The extension to higher dimensions, however, means that there are more degrees of freedom. First of all, the d-variate kernel has to be selected. Secondly, one has to decide on the particular smoothing parametrisation. A full bandwidth matrix allows for more flexibility; however, it also introduces more complexity into the estimator since more parameters need to be chosen. Lastly, once the smoothing parametrisation has been settled upon, the matrix **H** itself has to be chosen. Because of its role in controlling both the amount and direction of multivariate smoothing this choice is particularly important. We will take a closer look at each of these choices later in this chapter. But first we need some methodology for mathematical quantification of the performance of a multivariate kernel density estimator. This is developed in the following two sections.

4.3 Asymptotic MISE approximations

As in the univariate setting we are also able to obtain a simple asymptotic approximation to the MISE of a multivariate kernel density estimator under certain smoothness assumptions on the density f. These assumptions are needed to allow us to use a multivariate version of Taylor's theorem which we now state.

MULTIVARIATE TAYLOR'S THEOREM. *Let g be a d-variate function and $\boldsymbol{\alpha}_n$ be a sequence of $d \times 1$ vectors with all components tending to zero. Also, let $\mathcal{D}_g(\mathbf{x})$ be the vector of first-order partial derivatives of g and $\mathcal{H}_g(\mathbf{x})$ be the Hessian matrix of g, the $d \times d$ matrix having (i,j) entry equal to*

$$\frac{\partial^2}{\partial x_i \partial x_j} g(\mathbf{x}).$$

Then, assuming that all entries of $\mathcal{H}_g(\mathbf{x})$ are continuous in a neighbourhood of x, we have

$$g(\mathbf{x} + \boldsymbol{\alpha}_n) = g(\mathbf{x}) + \boldsymbol{\alpha}_n^T \mathcal{D}_g(\mathbf{x}) + \tfrac{1}{2}\boldsymbol{\alpha}_n^T \mathcal{H}_g(\mathbf{x})\boldsymbol{\alpha}_n + o(\boldsymbol{\alpha}_n^T \boldsymbol{\alpha}_n).$$

∎

For our asymptotic MISE approximations we will make the following assumptions on f, **H** and K:

4.3 ASYMPTOTIC MISE APPROXIMATIONS

(i) Each entry of $\mathcal{H}_f(\cdot)$ is piecewise continuous and square integrable.

(ii) $\mathbf{H} = \mathbf{H}_n$ is a sequence of bandwidth matrices such that $n^{-1}|\mathbf{H}|^{-1/2}$ and all entries of $\mathbf{H}$ approach zero as $n \to \infty$. Also, we assume that the ratio of the largest and smallest eigenvalues of $\mathbf{H}$ is bounded for all n.

(iii) K is a bounded, compactly supported d-variate kernel satisfying

$$\int K(\mathbf{z})\, d\mathbf{z} = 1 \quad \int \mathbf{z} K(\mathbf{z})\, d\mathbf{z} = 0$$

and

$$\int \mathbf{z}\mathbf{z}^T K(\mathbf{z})\, d\mathbf{z} = \mu_2(K)\mathbf{I}$$

where $\mu_2(K) = \int z_i^2 K(\mathbf{z})\, d\mathbf{z}$ is independent of i.

Note that condition (iii) is satisfied by all spherically symmetric compactly supported probability densities and product kernels based on compactly supported symmetric univariate densities. Also note that the assumption of compact support of K can be removed by imposing more complicated conditions on f.

The derivation of the asymptotic bias will require some additional matrix results. If $\mathbf{A}$ is a square matrix then the *trace* of $\mathbf{A}$, denoted by $\operatorname{tr} \mathbf{A}$, is the sum of the diagonal entries of $\mathbf{A}$ and has the property that

$$\operatorname{tr}(\mathbf{AB}) = \operatorname{tr}(\mathbf{BA}) \tag{4.3}$$

whenever both matrix products are defined. For a $d \times d$ matrix $\mathbf{A}$ the *vector* of $\mathbf{A}$, denoted by $\operatorname{vec} \mathbf{A}$, is the $d^2 \times 1$ vector obtained by stacking the columns of $\mathbf{A}$ underneath each other in order from left to right. The *vector-half* of $\mathbf{A}$, denoted by $\operatorname{vech} \mathbf{A}$, is the $\frac{1}{2}d(d+1) \times 1$ vector obtained from $\operatorname{vec} \mathbf{A}$ by eliminating each of the above-diagonal entries of $\mathbf{A}$ (Henderson and Searle, 1979).

EXAMPLE.

If $\mathbf{A} = \begin{bmatrix} 1 & 4 \\ 7 & 3 \end{bmatrix}$ then $\operatorname{vec} \mathbf{A} = \begin{bmatrix} 1 \\ 7 \\ 4 \\ 3 \end{bmatrix}$ and $\operatorname{vech} \mathbf{A} = \begin{bmatrix} 1 \\ 7 \\ 3 \end{bmatrix}$.

∎

Clearly, if $\mathbf{A}$ is symmetric then $\operatorname{vech} \mathbf{A}$ contains each of the distinct entries of $\mathbf{A}$. Since, for symmetric $\mathbf{A}$, $\operatorname{vec} \mathbf{A}$ contains the entries of vech $\mathbf{A}$ with some repetitions, there is a unique $d^2 \times \frac{1}{2}d(d+1)$ matrix $\mathbf{D}_d$ of zeros and ones such that

$$\mathbf{D}_d \text{vech } \mathbf{A} = \text{vec } \mathbf{A} \qquad (\mathbf{A} = \mathbf{A}^T) \qquad (4.4)$$

(see e.g. Magnus and Neudecker, 1988, p.49) and this is called the *duplication matrix* of order d. For example, the duplication matrix of order 2 is given by

$$\mathbf{D}_2 = \begin{bmatrix} 1 & 0 & 0 \\ 0 & 1 & 0 \\ 0 & 1 & 0 \\ 0 & 0 & 1 \end{bmatrix}.$$

A useful result that holds for all square matrices $\mathbf{A}$ is

$$\mathbf{D}_d^T \text{vec } \mathbf{A} = \text{vech } (\mathbf{A} + \mathbf{A}^T - \text{dg } \mathbf{A}), \qquad (4.5)$$

where $\text{dg } \mathbf{A}$ is the same as $\mathbf{A}$, but with all off-diagonal entries equal to zero. A further useful result is

$$\text{tr}(\mathbf{A}^T \mathbf{B}) = (\text{vec}^T \mathbf{A})(\text{vec } \mathbf{B}) \qquad (4.6)$$

(Exercise 4.2). Finally, for linear changes of variables when integrating over $\mathbb{R}^d$ we will need

$$\int g(\mathbf{A}\mathbf{x})\, d\mathbf{x} = |\mathbf{A}| \int g(\mathbf{y})\, d\mathbf{y}$$

where $\mathbf{A}$ represents an invertible $d \times d$ matrix.

We are now ready to derive the asymptotic bias of $\hat{f}(\mathbf{x}; \mathbf{H})$. By

4.3 ASYMPTOTIC MISE APPROXIMATIONS

the multivariate version of Taylor's theorem

$$\begin{aligned}
E\hat{f}(\mathbf{x};\mathbf{H}) &= \int K_{\mathbf{H}}(\mathbf{x}-\mathbf{y})f(\mathbf{y})\,d\mathbf{y} \\
&= \int K(\mathbf{z})f(\mathbf{x}-\mathbf{H}^{1/2}\mathbf{z})\,d\mathbf{z} \\
&= \int K(\mathbf{z})\{f(\mathbf{x}) - (\mathbf{H}^{1/2}\mathbf{z})^T \mathcal{D}_f(\mathbf{x}) \\
&\quad + \tfrac{1}{2}(\mathbf{H}^{1/2}\mathbf{z})^T \mathcal{H}_f(\mathbf{x})(\mathbf{H}^{1/2}\mathbf{z})\}\,d\mathbf{z} + o\{\mathrm{tr}(\mathbf{H})\} \\
&= f(\mathbf{x}) - \int \mathbf{z}^T \mathbf{H}^{1/2}\mathcal{D}_f(\mathbf{x})K(\mathbf{z})\,d\mathbf{z} \\
&\quad + \tfrac{1}{2}\int \mathbf{z}^T \mathbf{H}^{1/2}\mathcal{H}_f(\mathbf{x})\mathbf{H}^{1/2}\mathbf{z}K(\mathbf{z})\,d\mathbf{z} + o\{\mathrm{tr}(\mathbf{H})\} \\
&= f(\mathbf{x}) + \tfrac{1}{2}\mathrm{tr}\left\{\mathbf{H}^{1/2}\mathcal{H}_f(\mathbf{x})\mathbf{H}^{1/2}\int \mathbf{z}\mathbf{z}^T K(\mathbf{z})\,d\mathbf{z}\right\} \\
&\quad + o\{\mathrm{tr}(\mathbf{H})\} \\
&= f(\mathbf{x}) + \tfrac{1}{2}\mu_2(K)\mathrm{tr}\{\mathbf{H}\mathcal{H}_f(\mathbf{x})\} + o\{\mathrm{tr}(\mathbf{H})\}.
\end{aligned}$$

Therefore, the leading bias term is

$$E\hat{f}(\mathbf{x};\mathbf{H}) - f(\mathbf{x}) \sim \tfrac{1}{2}\mu_2(K)\mathrm{tr}\{\mathbf{H}\mathcal{H}_f(\mathbf{x})\}. \tag{4.7}$$

The variance of $\hat{f}(\mathbf{x};\mathbf{H})$ is given by

$$\begin{aligned}
\mathrm{Var}\hat{f}(\mathbf{x};\mathbf{H}) &= n^{-1}\bigg[|\mathbf{H}|^{-1/2}\int K(\mathbf{z})^2 f(\mathbf{x}-\mathbf{H}^{1/2}\mathbf{z})\,d\mathbf{z} \\
&\quad - \bigg\{\int K(\mathbf{z})f(\mathbf{x}-\mathbf{H}^{1/2}\mathbf{z})\bigg\}^2\bigg] \\
&= n^{-1}|\mathbf{H}|^{-1/2}R(K)f(\mathbf{x}) + o(n^{-1}|\mathbf{H}|^{-1/2}),
\end{aligned} \tag{4.8}$$

where $R(K) = \int K(\mathbf{z})^2\,d\mathbf{z}$. By our integrability assumptions in conditions (i) and (iii) we can combine these to obtain the AMISE of the multivariate kernel density estimator as

$$\begin{aligned}
\mathrm{AMISE}\{\hat{f}(\cdot;\mathbf{H})\} &= n^{-1}|\mathbf{H}|^{-1/2}R(K) \\
&\quad + \tfrac{1}{4}\mu_2(K)^2 \int \mathrm{tr}^2\{\mathbf{H}\mathcal{H}_f(\mathbf{x})\}\,d\mathbf{x}.
\end{aligned} \tag{4.9}$$

The integrated squared bias term can be expanded using (4.3) and (4.6):

$$\int \mathrm{tr}^2\{\mathbf{H}\mathcal{H}_f(\mathbf{x})\}\,d\mathbf{x}$$
$$= \int (\mathrm{vec}\,^T\mathbf{H})\{\mathrm{vec}\,\mathcal{H}_f(\mathbf{x})\}\{\mathrm{vec}\,^T\mathcal{H}_f(\mathbf{x})\}(\mathrm{vec}\,\mathbf{H})\,d\mathbf{x}$$
$$= \int (\mathrm{vech}\,^T\mathbf{H})\mathbf{D}_d^T\{\mathrm{vec}\,\mathcal{H}_f(\mathbf{x})\}\{\mathrm{vech}\,^T\mathcal{H}_f(\mathbf{x})\}\mathbf{D}_d(\mathrm{vech}\,\mathbf{H})\,d\mathbf{x}$$
$$= (\mathrm{vech}\,^T\mathbf{H})\,\boldsymbol{\Psi}_{\mathcal{F}}\,(\mathrm{vech}\,\mathbf{H})$$

where, from (4.5), $\boldsymbol{\Psi}_{\mathcal{F}}$ is the $\tfrac{1}{2}d(d+1)\times \tfrac{1}{2}d(d+1)$ matrix given by

$$\boldsymbol{\Psi}_{\mathcal{F}} = \int \mathrm{vech}\,\{2\mathcal{H}_f(\mathbf{x}) - \mathrm{dg}\,\mathcal{H}_f(\mathbf{x})\}$$
$$\times \mathrm{vech}\,^T\{2\mathcal{H}_f(\mathbf{x}) - \mathrm{dg}\,\mathcal{H}_f(\mathbf{x})\}\,d\mathbf{x}.$$

Thus, an alternative form for (4.9) is

$$\mathrm{AMISE}\{\hat{f}(\cdot;\mathbf{H})\} = n^{-1}|\mathbf{H}|^{-1/2}R(K) \qquad (4.10)$$
$$+ \tfrac{1}{4}\mu_2(K)^2(\mathrm{vech}\,^T\mathbf{H})\boldsymbol{\Psi}_{\mathcal{F}}(\mathrm{vech}\,\mathbf{H}).$$

At first glance the matrix $\boldsymbol{\Psi}_{\mathcal{F}}$ looks quite complicated. However, for sufficiently smooth f simple formulae for the entries may be obtained using integration by parts. For a d-variate function g and vector $\mathbf{r} = (r_1, \ldots, r_d)$ of non-negative integers we will use the notation

$$g^{(\mathbf{r})}(\mathbf{x}) = \frac{\partial^{|\mathbf{r}|}}{\partial x_1^{r_1}\ldots\partial x_d^{r_d}}g(\mathbf{x}),$$

assuming that this derivative exists. The notation $|\mathbf{r}|$ is for the sum of the entries of $\mathbf{r}$, that is $|\mathbf{r}| = \sum_{i=1}^{d} r_i$. We can show (see Exercise 4.3) that

$$\int f^{(\mathbf{r})}(\mathbf{x})f^{(\mathbf{r}')}(\mathbf{x})\,d\mathbf{x} = (-1)^{|\mathbf{r}|}\int f^{(\mathbf{r}+\mathbf{r}')}(\mathbf{x})f(\mathbf{x})\,d\mathbf{x} \qquad (4.11)$$

if $|\mathbf{r}+\mathbf{r}'|$ is even, and 0 otherwise. It follows from this that each entry of $\boldsymbol{\Psi}_{\mathcal{F}}$ can be written in the form

$$\psi_{\mathbf{r}} = \int f^{(\mathbf{r})}(\mathbf{x})f(\mathbf{x})\,d\mathbf{x}$$

4.3 ASYMPTOTIC MISE APPROXIMATIONS

where $|\mathbf{r}|$ is even.

EXAMPLE. Consider the case $d = 2$. Then from (4.11) we have

$$\boldsymbol{\Psi}_{\mathcal{F}} = \begin{bmatrix} \psi_{4,0} & 2\psi_{3,1} & \psi_{2,2} \\ 2\psi_{3,1} & 4\psi_{2,2} & 2\psi_{1,3} \\ \psi_{2,2} & 2\psi_{1,3} & \psi_{0,4} \end{bmatrix}.$$

■

Unlike the univariate setting explicit expressions for the AMISE-optimal bandwidth matrix are not available in general and this quantity can only be obtained numerically (see Wand, 1992a).

Considerable simplification of (4.10) is possible in the case where $\mathbf{H} \in \mathcal{D}$ and $\mathbf{H} \in \mathcal{S}$. First of all, one can show that if

$$\mathbf{H} = \mathrm{diag}(h_1^2, \ldots, h_d^2) = \mathrm{diag}(\mathbf{h}^2)$$

then (4.10) reduces to

$$\mathrm{AMISE}\{\hat{f}(\cdot;\mathbf{H})\} = n^{-1}R(K)\left(\prod_{j=1}^{d} h_j\right)^{-1} \\ + \tfrac{1}{4}\mu_2(K)^2(\mathbf{h}^2)^T \boldsymbol{\Psi}_{\mathcal{D}}(\mathbf{h}^2) \quad (4.12)$$

where $\boldsymbol{\Psi}_{\mathcal{D}}$ is the $d \times d$ matrix having (i,j) entry equal to $\psi_{2\mathbf{e}_i + 2\mathbf{e}_j}$ and $\mathbf{e}_i$ is the d-tuple having 1 in the ith position and 0 elsewhere (see Exercise 4.5). In the case where $\mathbf{H} = h^2 \mathbf{I}$ we obtain

$$\mathrm{AMISE}\{\hat{f}(\cdot;\mathbf{H})\} = n^{-1}h^{-d}R(K) \\ + \tfrac{1}{4}h^4 \mu_2(K)^2 \int \{\nabla^2 f(\mathbf{x})\}^2 \, d\mathbf{x}.$$

where

$$\nabla^2 f(\mathbf{x}) = \sum_{i=1}^{d} (\partial^2/\partial x_i^2) f(\mathbf{x}).$$

In this case the optimal bandwidth has an explicit formula and is given by

$$h_{\mathrm{AMISE}} = \left[\frac{dR(K)}{\mu_2(K)^2 \int \{\nabla^2 f(\mathbf{x})\}^2 \, d\mathbf{x}\, n}\right]^{1/(d+4)}$$

and the minimum AMISE is

$$\inf_{h>0} \text{AMISE}\{\hat{f}(\cdot;h)\}$$

$$= \frac{d+4}{4d} \left(\mu_2(K)^{2d}\{dR(K)\}^4 \left[\int \{\nabla^2 f(\mathbf{x})\}^2 \, d\mathbf{x} \right]^d n^{-4} \right)^{1/(d+4)}.$$

Notice that the rate of convergence of $\inf_{h>0} \text{MISE}\{\hat{f}(\cdot;h)\}$ is of order $n^{-4/(d+4)}$, a rate which becomes slower as the dimension increases. This slower rate is a manifestation of the curse of dimensionality as discussed previously and can severely inhibit the practical performance of density estimators in higher dimensions. However, there is substantial evidence that kernel density estimates are useful tools for recovering structure in moderate-dimensional data sets (Scott and Wand, 1991, Scott, 1992).

Finally, we comment on the evaluation of $\text{AMISE}\{\hat{f}(\cdot;\mathbf{H})\}$ for particular cases. It is useful to be able to do this for a wide class of densities so that a better understanding of the performance of multivariate density estimators can be gained. The comparison of smoothing parametrisations presented in Section 4.6 relies on such calculations. The main problem with (4.10) is that it involves multivariate integrals, and for many densities these could be difficult to evaluate, so a wide class of densities for which the $\psi_\mathbf{r}$ integrals have closed form is useful. The family of multivariate normal mixture densities satisfies this requirement. Our notation for the d-variate $N(\mathbf{0}, \boldsymbol{\Sigma})$ density is

$$\phi_{\boldsymbol{\Sigma}}(\mathbf{x}) = (2\pi)^{-d/2} |\boldsymbol{\Sigma}|^{-1/2} \exp(-\tfrac{1}{2}\mathbf{x}^T \boldsymbol{\Sigma}^{-1} \mathbf{x}).$$

It should be noted that this notation is not quite the extension of the univariate notation used in Chapter 2 since in the one dimensional case $\boldsymbol{\Sigma} = \sigma^2$ where σ^2 denotes the variance of the univariate normal distribution. In our univariate notation the subscript is σ rather than σ^2.

The family of multivariate normal mixture densities is given by

$$f(\mathbf{x}) = \sum_{\ell=1}^{k} w_\ell \phi_{\boldsymbol{\Sigma}_\ell}(\mathbf{x} - \boldsymbol{\mu}_\ell) \qquad (4.13)$$

where $\mathbf{w} = (w_1, \ldots, w_k)^T$ is a vector of positive numbers summing to one, and for each $\ell = 1, \ldots, k$, $\boldsymbol{\mu}_\ell$ is a $d \times 1$ vector and $\boldsymbol{\Sigma}_\ell$ is a $d \times d$ covariance matrix. For each pair (ℓ, ℓ') let

$$\mathbf{A}_{\ell\ell'} = (\boldsymbol{\Sigma}_\ell + \boldsymbol{\Sigma}_{\ell'})^{-1},$$

and
$$\mathbf{B}_{\ell\ell'} = \mathbf{A}_{\ell\ell'}\{\mathbf{I} - 2(\boldsymbol{\mu}_\ell - \boldsymbol{\mu}_{\ell'})(\boldsymbol{\mu}_\ell - \boldsymbol{\mu}_{\ell'})^T \mathbf{A}_{\ell\ell'}\}$$
$$\mathbf{C}_{\ell\ell'} = \mathbf{A}_{\ell\ell'}\{\mathbf{I} - (\boldsymbol{\mu}_\ell - \boldsymbol{\mu}_{\ell'})(\boldsymbol{\mu}_\ell - \boldsymbol{\mu}_{\ell'})^T \mathbf{A}_{\ell\ell'}\}.$$

Then we have

$$\text{AMISE}\{\hat{f}(\cdot;\mathbf{H})\} = n^{-1}R(K)|\mathbf{H}|^{-1/2} + \tfrac{1}{4}\mu_2(K)^2 \mathbf{w}^T \Xi \mathbf{w} \qquad (4.14)$$

where Ξ is the $k \times k$ matrix having (ℓ, ℓ') entry equal to

$$\phi_{\Sigma_\ell + \Sigma_{\ell'}}(\boldsymbol{\mu}_\ell - \boldsymbol{\mu}_{\ell'})\{2\text{tr}(\mathbf{HA}_{\ell\ell'}\mathbf{HB}_{\ell\ell'}) + \text{tr}^2(\mathbf{HC}_{\ell\ell'})\}$$

(Wand, 1992a).

4.4 Exact MISE calculations

In Section 2.6 we showed that explicit MISE expressions are available for the univariate kernel density estimator when f is a normal mixture density and K is a Gaussian kernel. These expressions provide a wide class of example densities for which one can perform exact MISE calculations without having to resort to numerical integration. In the multivariate setting the difficulties associated with numerical integration are magnified, so a flexible class of multivariate densities exhibiting explicit MISE expressions is useful to have. In this section we show that multivariate normal mixture densities serve this same purpose in the multivariate setting.

To allow explicit calculations for the multivariate case it is simplest to take $K = \phi_\mathbf{I}$, the d-variate normal kernel. For this kernel $\mu_2(K) = 1$ and $R(K) = (4\pi)^{-d/2}$. Recall that $K_\mathbf{H}$ is simply the $N(\mathbf{0}, \mathbf{H})$ density. The extension of (2.15) which we need for multivariate MISE formulae is

$$\int \phi_\Sigma(\mathbf{x} - \boldsymbol{\mu})\phi_{\Sigma'}(\mathbf{x} - \boldsymbol{\mu}')\,d\mathbf{x} = \phi_{\Sigma+\Sigma'}(\boldsymbol{\mu} - \boldsymbol{\mu}'). \qquad (4.15)$$

See Exercise 4.10 for the derivation of this result. In Section 2.6 the convolution notation was used to derive explicit MISE results. However, this notation is not essential and the exact MISE for estimation of (4.13) using the normal kernel can be derived by repeated application of (4.15).

First observe that

$$\mathrm{MISE}\{\hat{f}(\cdot;\mathbf{H})\} = \int \mathrm{Var}\{\hat{f}(\mathbf{x};\mathbf{H})\}\,d\mathbf{x}$$
$$+ \int \{E\hat{f}(\mathbf{x};\mathbf{H}) - f(\mathbf{x})\}^2\,d\mathbf{x}$$

and

$$\mathrm{Var}\{\hat{f}(\mathbf{x};\mathbf{H})\} = n^{-1}[E\phi_{\mathbf{H}}(\mathbf{x}-\mathbf{X})^2 - \{E\phi_{\mathbf{H}}(\mathbf{x}-\mathbf{X})\}^2].$$

From this we see explicit representation of $\mathrm{MISE}\{\hat{f}(\cdot;\mathbf{H})\}$ requires simplification of the integral $\int \{E\phi_{\mathbf{H}}(\mathbf{x}-\mathbf{X})\}^2\,d\mathbf{x}$, as well as several others. We will perform the required calculations for this integral; the others can be performed analogously. By (4.15) and symmetry of $\phi_{\mathbf{H}}$ we have

$$E\phi_{\mathbf{H}}(\mathbf{x}-\mathbf{X}) = \sum_{\ell=1}^{k} w_\ell \int \phi_{\mathbf{H}}(\mathbf{y}-\mathbf{x})\phi_{\Sigma_\ell}(\mathbf{y}-\boldsymbol{\mu}_\ell)\,d\mathbf{y}$$
$$= \sum_{\ell=1}^{k} w_\ell \phi_{\mathbf{H}+\Sigma_\ell}(\mathbf{x}-\boldsymbol{\mu}_\ell).$$

This leads to

$$\int \{E\phi_{\mathbf{H}}(\mathbf{x}-\mathbf{X})\}^2\,d\mathbf{x}$$
$$= \sum_{\ell=1}^{k}\sum_{\ell'=1}^{k} w_\ell w_{\ell'} \int \phi_{\mathbf{H}+\Sigma_\ell}(\mathbf{x}-\boldsymbol{\mu}_\ell)\phi_{\mathbf{H}+\Sigma_{\ell'}}(\mathbf{x}-\boldsymbol{\mu}_{\ell'})\,d\mathbf{x}$$
$$= \sum_{\ell=1}^{k}\sum_{\ell'=1}^{k} w_\ell w_{\ell'} \phi_{2\mathbf{H}+\Sigma_\ell+\Sigma_{\ell'}}(\boldsymbol{\mu}_\ell - \boldsymbol{\mu}_{\ell'}) = \mathbf{w}^T \boldsymbol{\Omega}_2 \mathbf{w}$$

where $\boldsymbol{\Omega}_a$ denotes the $k \times k$ matrix with (ℓ,ℓ') entry equal to

$$\phi_{a\mathbf{H}+\Sigma_\ell+\Sigma_{\ell'}}(\boldsymbol{\mu}_\ell - \boldsymbol{\mu}_{\ell'}).$$

The other terms can be handled in the same way to obtain

$$\mathrm{MISE}\{\hat{f}(\cdot;\mathbf{H})\} = n^{-1}(4\pi)^{-d/2}|\mathbf{H}|^{-1/2} \\ + \mathbf{w}^T\{(1-n^{-1})\boldsymbol{\Omega}_2 - 2\boldsymbol{\Omega}_1 + \boldsymbol{\Omega}_0\}\mathbf{w} \quad (4.16)$$

(Exercise 4.11) which is the multivariate extension of (2.18). This is a useful formula for finite sample analyses of multivariate kernel density estimators.

4.5 Choice of a multivariate kernel

Multivariate kernel functions can be constructed from univariate ones in various ways. As mentioned in Section 4.2 there are two popular ways of doing so when we wish the kernel to be a multivariate density itself; these are the product kernel $K^P(\mathbf{x}) = \prod_{i=1}^{d} \kappa(x_i)$ and the spherically symmetric kernel $K^S(\mathbf{x}) = c_{\kappa,d} \kappa\{(\mathbf{x}^T\mathbf{x})^{1/2}\}$ where κ is a symmetric univariate kernel density function. In this section, we shall make some theoretical comparisons between product and spherically symmetric kernels obtained from the same univariate kernel κ. In general K^P and K^S will be different, the exception being the normal density which is the only multivariate density that is both spherically symmetric and has independent marginals (e.g. Muirhead, 1982, p.35). For example, Figure 4.3 shows contour plots of (a) the bivariate product and (b) the bivariate spherically symmetric versions of the biweight kernel introduced in Section 2.7.

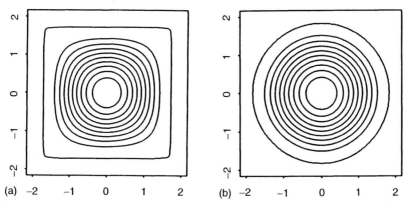

Figure 4.3. *Bivariate kernels based on the biweight kernel: (a) the product kernel and (b) the spherically symmetric kernel. Each kernel has been canonically scaled.*

To make the theoretical comparison between product and spherically symmetric kernels, we need the multivariate analogue of the quantity $C(K) = \{R(K)^4 \mu_2(K)^2\}^{1/5}$ that determined the AMISE performance of univariate kernels in Section 2.7. This can be shown to be

$$C_d(K) = \{R(K)^4 \mu_2(K)^{2d}\}^{1/(d+4)}$$

(Exercise 4.12). The efficiency of a spherically symmetric kernel K^S relative to the product kernel K^P will be measured by the

ratio
$$\{C_d(K^S)/C_d(K^P)\}^{(d+4)/4}$$
$$= \{R(K^S)\mu_2(K^S)^{d/2}\}/\{R(K^P)\mu_2(K^P)^{d/2}\}.$$

This is the d-variate extension of the ratio used in Table 2.1 to compare univariate kernels.

Table 4.1 gives values of this efficiency for kernels based on the univariate symmetric beta kernels discussed in Section 2.7, that is, those of the form

$$\kappa(x) \propto (1-x^2)^p 1_{\{|x|<1\}}.$$

Recall that values less than 1 indicate asymptotic superiority of the spherically symmetric extension of κ. We see that in the bivariate case there is little difference between product and spherically symmetric cases, even in the uniform case where there is only a 4.5% loss in efficiency by using the product extension. For three and four dimensions, the difference increases, and for the uniform kernels in particular the efficiencies become appreciably worse. There is less difference for the cases $p = 1$, 2 and 3.

Table 4.1. *Efficiencies of product relative to spherically symmetric kernels when the basic kernel is the symmetric beta with parameter p.*

p	$d=2$	$d=3$	$d=4$
0	0.955	0.888	0.811
1	0.982	0.953	0.916
2	0.983	0.953	0.915
3	0.984	0.956	0.919

Within each class of multivariate kernels, it is possible to compute the optimal kernel in the AMISE sense. For product kernels, the quantity to be minimised can be written as

$$C_d(K^P) = \{R(\kappa)^4 \mu_2(\kappa)^2\}^{d/(d+4)}.$$

The minimiser of this expression over κ corresponds to the minimiser of $C(K)$, namely the Epanechnikov kernel. Therefore, the AMISE-optimal product kernel is

$$K_*^P(\mathbf{x}) = (\tfrac{3}{4})^d \prod_{i=1}^{d}(1-x_i^2)1_{\{|x_i|<1\}}.$$

4.6 CHOICE OF SMOOTHING PARAMETRISATION

For spherically symmetric kernels the optimisation problem is slightly more complicated. However, it can be shown (see Fukunaga and Hostetler, 1975, Müller, 1988, pp. 82–83) that the Epanechnikov kernel also forms the basis of the optimum and the optimal spherically symmetric multivariate kernel can be shown to be

$$K_*^S(\mathbf{x}) = \tfrac{1}{2} v_d^{-1}(d+2)(1 - \mathbf{x}^T\mathbf{x}) 1_{\{\mathbf{x}^T\mathbf{x} \leq 1\}},$$

where $v_d = 2\pi^{d/2}/\{d\Gamma(d/2)\}$ is the volume of the unit d-dimensional sphere.

4.6 Choice of smoothing parametrisation

As we saw in Section 4.2, there are several levels of sophistication when specifying the bandwidth matrix $\mathbf{H}$. The simplest corresponds to the restriction $\mathbf{H} \in \mathcal{S}$ which means that $\mathbf{H} = h^2 \mathbf{I}$ for some $h > 0$. This restriction has the advantage that one only has to deal with a single smoothing parameter, but the considerable disadvantage that the amount of smoothing is the same in each coordinate direction. At the next level, $\mathbf{H} \in \mathcal{D}$, $\mathbf{H} = \text{diag}(h_1^2, \ldots, h_d^2)$, so at the expense of introducing $d - 1$ additional smoothing parameters, one has the flexibility to smooth by different amounts in each of the d coordinate directions. However, there are situations where one might wish to smooth in directions different to those of the coordinate axes. In this case the full bandwidth matrix, $\mathbf{H} \in \mathcal{F}$, would be appropriate. These ideas are most easily understood by focusing on the bivariate context.

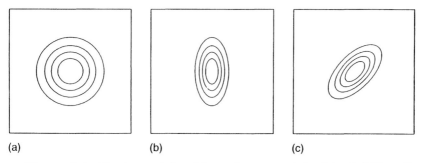

(a)   (b)   (c)

Figure 4.4. *Contour plots of kernels parametrised by (a)* $\mathbf{H} \in \mathcal{S}$, *(b)* $\mathbf{H} \in \mathcal{D}$ *and (c)* $\mathbf{H} \in \mathcal{F}$.

In Figure 4.4 contour plots of the Gaussian kernel parametrised by **H** being a typical member of each class are given. The kernel contours in each case are elliptical; however, when $\mathbf{H} \in \mathcal{S}$ these ellipses are constrained to be circles. For $\mathbf{H} \in \mathcal{D}$ the ellipses are such that their axes correspond to the coordinate directions. It is only in the full matrix case that ellipses of arbitrary orientation are obtainable. This raises the question of how much can be gained from the more sophisticated smoothing parametrisations, keeping in mind that one has to deal with more parameters, with obvious difficulties for subjective choice of bandwidths, and repercussions for difficulty of automatic bandwidth selection.

Before attempting to answer this question we should first mention one simple way of obtaining a bandwidth matrix of arbitrary orientation. This involves taking **H** to be of the form

$$\mathbf{H} = h^2 \mathbf{S}$$

where **S** is the sample covariance matrix (Fukunaga, 1972, Silverman, 1986). This approach is equivalent to linearly transforming the data to have unit covariance matrix, often called *sphering* the data, applying the simple kernel estimator (4.2) to the sphered data and then "backtransforming" to obtain the density estimator of the original data. It can be shown (Exercise 4.7) that if f is the multivariate $N(\boldsymbol{\mu}, \boldsymbol{\Sigma})$ distribution then the AMISE-optimal **H** satisfies

$$\mathbf{H}_{\text{AMISE}} = c\boldsymbol{\Sigma} \tag{4.17}$$

for a scalar constant c depending only on d and n. This result shows that, in the case of multivariate normal data, sphering is appropriate. However, there is no corresponding theoretical support for sphering for estimation of general density shapes.

One way of comparing the cost due to using a particular class of bandwidth matrix when estimating a given density f is through an asymptotic relative efficiency (ARE) criterion. For a particular bandwidth matrix class $\mathcal{A}$, the ARE of $\mathcal{A}$ compared to the full class $\mathcal{F}$ is given by

$$\text{ARE}_f(\mathcal{F} : \mathcal{A})$$
$$= \left\{ \inf_{\mathbf{H} \in \mathcal{F}} \text{AMISE}(\hat{f}(\cdot; \mathbf{H})) / \inf_{\mathbf{H} \in \mathcal{A}} \text{AMISE}(\hat{f}(\cdot; \mathbf{H})) \right\}^{(d+4)/4}$$

Since, over each class, $\inf_{\mathbf{H} \in \mathcal{A}} \text{AMISE}(\hat{f}(\cdot; \mathbf{H})) = O(n^{-4/(d+4)})$, this quantity has the interpretation that a sample of size n using

4.6 CHOICE OF SMOOTHING PARAMETRISATION

H from $\mathcal{A}$ has the same minimum AMISE as a sample of size $\text{ARE}_f(\mathcal{F} : \mathcal{A})n$ when using the full **H** matrix.

It is of practical relevance to compare the performance of the smaller bandwidth matrix classes to the full class $\mathcal{F}$. Obviously there will be some loss of efficiency due to using fewer parameters, but of interest is just how large this loss can be. This depends on the particular density shape and orientation with respect to the coordinate axes, so a fuller study would take up more space than we can devote here. However, we can get some feeling from a few simple examples. Figure 4.5 shows contour plots of two densities, the first is a bivariate normal centred at the origin with variances 1 and $\frac{1}{4}$ and zero covariance. The second is an equal proportion mixture of two normals having means $(-\frac{3}{2}, 0)^T$ and $(\frac{3}{2}, 0)^T$, variances $(\frac{1}{16})^2$ and 1 and zero covariance.

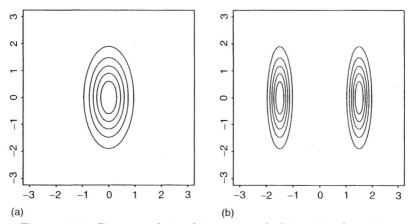

Figure 4.5. *Contour plots of two example bivariate densities*

There will be no loss in efficiency due to using $\mathbf{H} \in \mathcal{D}$ compared to $\mathbf{H} \in \mathcal{F}$ since the probability mass happens to correspond to the coordinate axes. If the probability mass undergoes some rotation then this is no longer the case and we should expect the efficiency to fall below 1. In Figure 4.6 we have plotted values of $\text{ARE}_f(\mathcal{F} : \mathcal{D})$ versus θ, where θ is the angle of rotation ($0 \leq \theta \leq \pi$) of the probability mass about the origin. The dashed line in each picture corresponds to the ARE of the sphering approach. In view of (4.17) it comes as no surprise that the sphering approach has ARE equal to 1 for the bivariate normal case. However, for a simple non-Gaussian density sphering can be very detrimental, with an ARE of 0.26 for the second density shape. For this density the variance of the horizontal coordinate variable is a very poor surrogate for measuring the optimal amount of smoothing in the horizontal direction because it does not take into account the "within-modes" curvature. One could, in fact, make the ARE of the sphering approach arbitrarily small by taking the modes to be sufficiently far apart since this increases the variance, but not the optimal amount of smoothing.

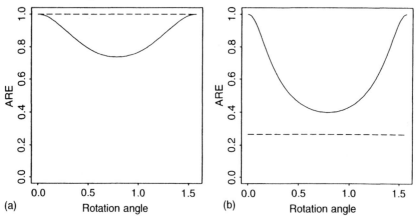

Figure 4.6. *Plots of* $\text{ARE}_f(\mathcal{F} : \mathcal{D})$ *versus angle of rotation* θ *for (a) bivariate normal density, (b) two-component normal mixture density.*

Even from this small number of examples we see that the choice of smoothing parametrisation can have a significant influence on the performance of a multivariate kernel estimate (see Wand and Jones, 1993, for a more complete study). Broadly speaking, one often does not lose very much by using a diagonal bandwidth matrix, although full matrices are necessary in some circumstances. However, the other simplified parametrisations have too many failings to be recommended in general.

4.7 Bandwidth selection

At the time of writing, the problem of selecting a bandwidth matrix from the data had received considerably less attention in the literature that its univariate counterpart. However, many of the ideas discussed in Chapter 3 for selecting h can be extended to the multivariate case. In this section we will briefly discuss some of these ideas, without getting into the more practical issues.

It is fairly obvious to see how one would generalise the least squares cross-validation selector described in Section 3.3 to allow

4.7 BANDWIDTH SELECTION

for selection of a bandwidth matrix $\mathbf{H}$. The LSCV selector of $\mathbf{H}$ is

$$\hat{\mathbf{H}}_{\text{LSCV}} = \text{argmin}_{\mathbf{H} \in \mathcal{F}} \text{LSCV}(\mathbf{H})$$

where

$$\text{LSCV}(\mathbf{H}) = \int \hat{f}(\mathbf{x}; \mathbf{H})^2 \, d\mathbf{x} - 2n^{-1} \sum_{i=1}^{n} \hat{f}_{-i}(\mathbf{X}_i; \mathbf{H})$$

and $\hat{f}_{-i}(\cdot; \mathbf{H})$ is the kernel estimator based on the sample with $\mathbf{X}_i$ deleted. Of course, one can also minimise LSCV($\mathbf{H}$) over $\mathbf{H} \in \mathcal{D}$ or $\mathbf{H} \in \mathcal{S}$ to obtain bandwidth selectors belonging to the smaller classes. The unfortunate theoretical aspects of univariate least squares cross-validation, as discussed in Section 3.7, are also present in the multivariate setting. However, somewhat surprisingly, the relative rate of convergence of LSCV improves for higher dimensions (Hall and Marron, 1987a). Biased cross-validation also has a straightforward extension to higher dimensions (Sain, Baggerly and Scott, 1994).

Given the more stable performance of plug-in approaches in the univariate setting it seems worthwhile to investigate the performance of their multivariate extensions. Using the asymptotic approximations developed in Section 4.3 it is possible to develop multivariate versions of plug-in type bandwidth selectors. From (4.10) and the discussion following we see that the only unknown quantities are of the form

$$\psi_{\mathbf{r}} = \int f^{(\mathbf{r})}(\mathbf{x}) f(\mathbf{x}) \, d\mathbf{x}.$$

This functional can be replaced by a kernel estimator of the form

$$\hat{\psi}_{\mathbf{r}}(\mathbf{G}) = n^{-1} \sum_{i=1}^{n} \hat{f}^{(\mathbf{r})}(\mathbf{X}_i; \mathbf{G}) \quad (4.18)$$

where $\mathbf{G}$ is another bandwidth matrix. This allows one to plug in estimates of the $\psi_{\mathbf{r}}$ and minimise the resulting estimate of the AMISE (Wand and Jones, 1994). The asymptotic mean squared error properties of $\hat{\psi}_{\mathbf{r}}(\mathbf{G})$ are discussed in Exercise 4.13.

While it is clear that several possibilities exist for data driven bandwidth selection in the multivariate case, many of the practical issues are yet to be resolved. For this reason we will not discuss them here any further.

4.8 Bibliographical notes

4.2 The multivariate kernel density estimator was first considered in single bandwidth form by Cacoullos (1966) and with a vector of bandwidths by Epanechnikov (1969). Deheuvels (1977b) was the first to treat the full bandwidth matrix case. Singh (1976) extended these ideas to the estimation of partial derivatives of a multivariate density. For recent studies of the feasibility of multivariate density estimation see Scott and Wand (1991) and Terrell and Scott (1992). Scott (1992) provided a detailed study of the practical implementation and display of multivariate density estimates. Friedman, Stuetzle and Schroeder (1984) used univariate kernel estimators to construct a multivariate density estimate based on projection pursuit ideas. Multivariate kernel estimators have also been discussed in the monographs of Prakasa Rao (1983) and Silverman (1986).

4.3 A recent study of the properties of the kernel density estimator with full bandwidth matrix was carried out by Wand (1992a).

4.4 Exact multivariate MISE calculations were derived in Wand and Jones (1994).

4.6 This section is based on part of a detailed comparison of bandwidth parametrisations by Wand and Jones (1993). Fukunaga (1972) proposed the sphering idea for multivariate density estimation; see also Silverman (1986).

4.7 Relevant references for bandwidth selection in the multivariate case include Stone (1984), Terrell (1990), Sain, Baggerly and Scott (1994) and Wand and Jones (1994).

4.9 Exercises

4.1 To get a feeling for the "curse of dimensionality", suppose data arise from a uniform distribution on $[-1,1]^d$, and consider the kernel density estimator of $f(\mathbf{0})$, where $\mathbf{0}$ is the origin. Suppose that the kernel is a product kernel based on the uniform univariate kernel having support on $[-1,1]$ and that $\mathbf{H} = h^2\mathbf{I}$. Derive an expression for the expected proportion of points included within the kernel's support for general h (≤ 1) and d. Tabulate values of this expression for $h = 0.1, 0.5, 0.9$ and $d = 1, 2, 5, 10$ and 25.

4.2 Let $\mathbf{A}$ and $\mathbf{B}$ be $d \times d$ matrices.
(a) Show that $\operatorname{tr}(\mathbf{AB}) = \operatorname{tr}(\mathbf{BA})$.
(b) Show that $\operatorname{tr}(\mathbf{A}^T\mathbf{B}) = (\operatorname{vec}^T\mathbf{A})(\operatorname{vec}\mathbf{B})$.
(c) Verify (4.4) and (4.5) in the case $d = 2$.

4.3
(a) Suppose that u and v are functions on $\mathbb{R}^d$ such that

$$\lim_{|x_i|\to\infty} u^{(\mathbf{j})}(\mathbf{x}) = \lim_{|x_i|\to\infty} v^{(\mathbf{j})}(\mathbf{x}) = 0$$

for all $\mathbf{j} = (j_1, \ldots, j_d)$ such that $0 \leq j_i < r_i - 1$ and all $i = 1, \ldots, d$. Show that

$$\int u^{(\mathbf{r})}(\mathbf{x})v(\mathbf{x})\, d\mathbf{x} = (-1)^{|\mathbf{r}|} \int u(\mathbf{x})v^{(\mathbf{r})}(\mathbf{x})\, d\mathbf{x}$$

where $\mathbf{r} = (r_1, \ldots, r_d)$. Assume that all stated derivatives of u and v exist.

(b) Use the result in (a) to derive (4.11).

4.4 Find an expression for $\boldsymbol{\Psi}_{\mathcal{F}}$ in terms of $\psi_{\mathbf{r}}$ functionals for the case $d = 3$.

4.5 Show that (4.12) follows from (4.10) when $\mathbf{H} \in \mathcal{D}$.

4.6 Suppose that $d = 2$ and that the bandwidth matrix is of the form $\mathbf{H} = \operatorname{diag}(h_1^2, h_2^2)$. Show that the AMISE-optimal bandwidths are

$$h_{1,\text{AMISE}} = \left[\frac{\psi_{0,4}^{3/4} R(K)}{\mu_2(K)^2 \psi_{4,0}^{3/4} \{\psi_{2,2} + \psi_{0,4}^{1/2}\psi_{4,0}^{1/2}\}n}\right]^{1/6}$$

and

$$h_{2,\text{AMISE}} = \left[\frac{\psi_{4,0}^{3/4} R(K)}{\mu_2(K)^2 \psi_{0,4}^{3/4} \{\psi_{2,2} + \psi_{0,4}^{1/2}\psi_{4,0}^{1/2}\}n}\right]^{1/6}.$$

4.7 Suppose that f is the d-variate $N(\boldsymbol{\mu}, \boldsymbol{\Sigma})$ density and $K = \phi_\mathbf{I}$ is the d-variate Gaussian kernel.

(a) Using (4.14) show that

$$\text{AMISE}\{\hat{f}(\cdot;\mathbf{H})\} = n^{-1}(4\pi)^{-d/2}|\mathbf{H}|^{-1/2}$$
$$+ \tfrac{1}{16}(4\pi)^{-d/2}|\boldsymbol{\Sigma}|^{-1/2}\{2\operatorname{tr}(\mathbf{H}\boldsymbol{\Sigma}^{-1}\mathbf{H}\boldsymbol{\Sigma}^{-1}) + \operatorname{tr}^2(\mathbf{H}\boldsymbol{\Sigma}^{-1})\}.$$

(b) It can be shown that the bandwidth matrix that minimises this expression is

$$\mathbf{H}_{\text{AMISE}} = \{4/(d+2)\}^{2/(d+4)} \boldsymbol{\Sigma} n^{-2/(d+4)}.$$

Show that

$$\inf_{\mathbf{H}\in\mathcal{F}} \mathrm{AMISE}\{\hat{f}(\cdot;\mathbf{H})\}$$
$$= (4\pi)^{-d/2}\left(\frac{d+4}{4}\right)\left(\frac{d+2}{4}\right)^{d/(d+4)} |\Sigma|^{-1/2} n^{-4/(d+4)}.$$

4.8 Suppose that f is a bivariate normal density with correlation coefficient ρ and $K = \phi_{\mathbf{I}}$.
(a) Show that

$$\mathrm{ARE}_f(\mathcal{F}:\mathcal{D}) = \{2(1-\rho^2)/(2+\rho^2)\}^{1/2}.$$

(b) Determine the value of $\lim_{\rho\to 1} \mathrm{ARE}_f(\mathcal{F}:\mathcal{D})$.

4.9 Consider the kernel estimator of the order $\mathbf{r}$ partial derivative of f, $f^{(\mathbf{r})}$, based on the normal kernel

$$\hat{f}^{(\mathbf{r})}(\mathbf{x};\mathbf{H}) = n^{-1}\sum_{i=1}^{n}\phi_{\mathbf{H}}^{(\mathbf{r})}(\mathbf{x}-\mathbf{X}_i).$$

Show that if $f^{(\mathbf{r})}$ has continuous and integrable second-order partial derivatives then

$$\mathrm{AMISE}\{\hat{f}^{(\mathbf{r})}(\cdot;\mathbf{H})\} = (-1)^{|\mathbf{r}|} n^{-1}\phi_{2\mathbf{H}}^{(2\mathbf{r})}(\mathbf{0})$$
$$+ \tfrac{1}{4}\int \mathrm{tr}^2\{\mathbf{H}\mathcal{H}_{f^{(\mathbf{r})}}(\mathbf{x})\}\, d\mathbf{x}.$$

Simplify this expression as much as possible in the case where $d=2$ and $\mathbf{r}=(1,0)$. (Hint: results in Appendix C may be useful.)

4.10
(a) Show that for any two multivariate normal distributions $N(\boldsymbol{\mu},\boldsymbol{\Sigma})$ and $N(\boldsymbol{\mu}',\boldsymbol{\Sigma}')$

$$\phi_{\boldsymbol{\Sigma}}(\mathbf{x}-\boldsymbol{\mu})\phi_{\boldsymbol{\Sigma}'}(\mathbf{x}-\boldsymbol{\mu}')$$
$$= \phi_{\boldsymbol{\Sigma}+\boldsymbol{\Sigma}'}(\boldsymbol{\mu}-\boldsymbol{\mu}')\phi_{\boldsymbol{\Sigma}(\boldsymbol{\Sigma}+\boldsymbol{\Sigma}')^{-1}\boldsymbol{\Sigma}'}(\mathbf{x}-\boldsymbol{\mu}^*)$$

where
$$\boldsymbol{\mu}^* = \boldsymbol{\Sigma}'(\boldsymbol{\Sigma}+\boldsymbol{\Sigma}')^{-1}\boldsymbol{\mu} + \boldsymbol{\Sigma}(\boldsymbol{\Sigma}+\boldsymbol{\Sigma}')^{-1}\boldsymbol{\mu}'.$$

(b) Use this result to derive (4.15).

4.11 Derive expression (4.16) for a general normal mixture density.

4.12 By considering the scaling $K_{\boldsymbol{\Delta}}$ when $\boldsymbol{\Delta} = \delta^2 \mathbf{I}$, obtain a multivariate version of canonical scaling (Section 2.7). In particular, show that the multivariate analogue of the univariate quantity $C(K) = \{R(K)^4 \mu_2(K)^2\}^{1/5}$ is

$$C_d(K) = \{R(K)^4 \mu_2(K)^{2d}\}^{1/(d+4)}.$$

4.13 Consider the kernel functional estimator $\hat{\psi}_{\mathbf{r}}(\mathbf{G})$ defined at (4.18).

(a) Using techniques similar to those used to derive AMISE expressions in Section 4.3 show that

$$E\hat{\psi}_{\mathbf{r}}(\mathbf{G}) - \psi_{\mathbf{r}} \sim n^{-1} K_{\mathbf{G}}^{(\mathbf{r})}(\mathbf{0})$$
$$+ \tfrac{1}{2}\mu_2(K) \int \mathrm{tr}\{\mathbf{G}\mathcal{H}_f(\mathbf{x})\} f^{(\mathbf{r})}(\mathbf{x})\, d\mathbf{x}$$

and

$$\mathrm{Var}\{\hat{\psi}_{\mathbf{r}}(\mathbf{G})\} \sim 2n^{-2}\psi_{\mathbf{0}} \int K_{\mathbf{G}}^{(\mathbf{r})}(\mathbf{x})^2\, d\mathbf{x}$$
$$+ 4n^{-1}\left\{\int f^{(\mathbf{r})}(\mathbf{x})^2 f(\mathbf{x})\, d\mathbf{x} - \psi_{\mathbf{r}}^2\right\}.$$

(b) Suppose that $\mathbf{G} = g^2 \mathbf{I}$ and $K^{(\mathbf{r})}(\mathbf{0}) \neq 0$. Show that the leading bias terms cancel when $g = g_{\mathrm{AMSE}}$, where

$$g_{\mathrm{AMSE}} = \left[\frac{-2K^{(\mathbf{r})}(\mathbf{0})}{\mu_2(K)\left(\sum_{i=1}^{d} \psi_{\mathbf{r}+2\mathbf{e}_i}\right)n}\right]^{1/(2+d+|\mathbf{r}|)}.$$

CHAPTER 5

Kernel regression

5.1 Introduction

We are now in a position to return to the nonparametric regression problem introduced in Chapter 1. The main ideas and mathematical skills developed in the analysis of the kernel density estimator carry over to the kernel regression context for both univariate and multivariate data.

There now exist several approaches to the nonparametric regression problem. Some of the more popular are those based on kernel functions, spline functions and wavelets. Each of these approaches has its own particular strengths and weaknesses, although kernel regression estimators have the advantage of mathematical and intuitive simplicity. Within each of these broad classes of nonparametric regression estimators there are also several different approaches. In the context of kernel regression traditional approaches have involved the *Nadaraya-Watson estimator* (Nadaraya, 1964, Watson, 1964) and some alternative kernel estimators (Priestley and Chao, 1972, Gasser and Müller, 1979).

In this chapter we will study a class of kernel-type regression estimators called *local polynomial kernel estimators* (Stone, 1977, Cleveland, 1979, Müller, 1987, Fan, 1992a). These estimate the regression function at a particular point by "locally" fitting a pth degree polynomial to the data via weighted least squares. This class includes, as a special case, the Nadaraya-Watson estimator since it can be shown to correspond to fitting degree zero polynomials, that is, local constants. Of particular importance and simplicity is the *local linear kernel estimator*, corresponding to $p = 1$. The local linear kernel estimator also shares some similarities with the above-mentioned traditional kernel regression estimators although it has favourable asymptotic properties and boundary behaviour compared with those. We will also see that the mean squared

5.1 INTRODUCTION

error properties of the local linear kernel regression estimator are analogous to those of the kernel density estimator. This means that most of the ideas developed in the density estimation context can be easily transported to the regression context.

Before commencing our study of kernel-based nonparametric regression we will give some relevant terminology and notation. Nonparametric regression is studied in both *fixed design* and *random design* contexts. In the univariate fixed design case the design consists of $x_1, \ldots, x_n$ which are ordered non-random numbers. An *equally spaced* fixed design is one for which $x_{i+1} - x_i$ is constant for all i. For the fixed design case the response variables are assumed to satisfy

$$Y_i = m(x_i) + v^{1/2}(x_i)\varepsilon_i, \qquad i = 1, \ldots, n$$

where $\varepsilon_1, \ldots, \varepsilon_n$ are independent random variables for which

$$E(\varepsilon_i) = 0 \quad \text{and} \quad \text{Var}(\varepsilon_i) = 1.$$

We call m the *mean regression function*, or simply the *regression function*, since $E(Y_i) = m(x_i)$, while v is called the *variance function* since $\text{Var}(Y_i) = v(x_i)$. Often it is assumed that $v(x_i) = \sigma^2$ for all i, in which case the model is said to be *homoscedastic*. Otherwise the model is *heteroscedastic*.

The random design regression model arises when we observe a bivariate sample $(X_1, Y_1), \ldots, (X_n, Y_n)$ of random pairs in which case the model can be written as

$$Y_i = m(X_i) + v^{1/2}(X_i)\varepsilon_i, \qquad i = 1, \ldots, n$$

where, conditional on $X_1, \ldots, X_n$, the ε_i are independent random variables with zero mean and unit variance. However, in the random design context

$$m(x) = E(Y|X = x) \quad \text{and} \quad v(x) = \text{Var}(Y|X = x)$$

are, respectively, the conditional mean and variance of Y given $X = x$. The density of $X_1, \ldots, X_n$ will be denoted throughout by f.

5.2 Local polynomial kernel estimators

The local linear kernel estimator was introduced in Section 1.1 as a nonparametric estimate for the age/incomes data. In Figure 5.1 we present a further example that again shows how the local linear kernel estimator is constructed. The true regression function is

$$m(x) = 2\exp[-x^2/\{2(0.3)^2\}] + 3\exp[-(x-1)^2/\{2(0.7)^2\}] \quad (5.1)$$

confined to the interval $[0, 1]$, and is represented by the dashed curve. The data $Y_1, \ldots, Y_n$ were generated using

$$Y_i = m(x_i) + 0.075\varepsilon_i, \quad i = 1, \ldots, 100$$

where $x_i = i/100$ and the ε_i are independent $N(0,1)$ random variables. The (x_i, Y_i) pairs are represented by the crosses.

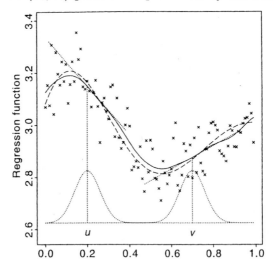

Figure 5.1. *Local linear kernel estimate (solid curve) of the regression function m given by (5.1) based on 100 simulated observations (represented by crosses). The dashed curve is the true function m. The dotted curves are the kernel weights and linear fits at the points u and v.*

The local linear kernel estimate corresponds to the solid line and the dotted lines show how it is constructed at two different points. At each of these points the dotted line was obtained by fitting a straight line to the Y_i using *weighted* least squares where the weights are chosen according to the height of the kernel function

5.2 LOCAL POLYNOMIAL KERNEL ESTIMATORS

centred about that point, and shown by the dotted curve at the base of the picture. When this local fitting process is performed at each point $x \in [0, 1]$ then the solid curve results. If K_h is, again, a kernel function scaled by a bandwidth h then, for estimation at a particular x, the weight assigned to a particular point Y_i is $K_h(x_i - x)$. Given the usual kernel shape this means that those observations close to x have more influence on the regression estimate at x than those farther away. The amount of relative influence is controlled by the bandwidth h which plays an analogous role to that for the kernel density estimator. If h is small the local linear fitting process depends heavily on those observations that are closest to x and tends to yield a more wiggly estimate. As h becomes closer to zero the estimator tends towards interpolation of the data. This is shown in Figure 5.2 (a) where a very small bandwidth is used. On the other hand, larger h tends to weight the observations more equally and as h increases the estimate tends towards the least squares line through the data. The estimate in Figure 5.2 (b) was obtained using a very large bandwidth and is indistinguishable from the least squares line over the range of the data.

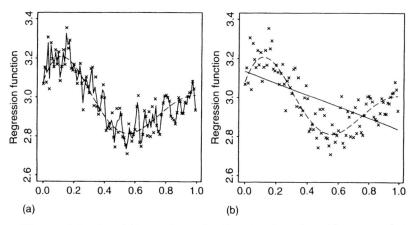

Figure 5.2. *Local linear kernel estimates based on the same data as used in Figure 5.1, but with (a) a very small bandwidth (b) a very large bandwidth.*

A natural extension of the local linear kernel estimator is that which fits higher degree polynomials locally. Figure 5.3 shows a local cubic fit to the data with the dotted lines showing the cubic polynomial fits at locations u and v. Notice that the peak and valley of m are better estimated by the local cubic fit since cubics have more degrees of freedom for estimating regions of high curvature than do lines, although this estimate is more computationally complex and suffers from a higher degree of sample variability.

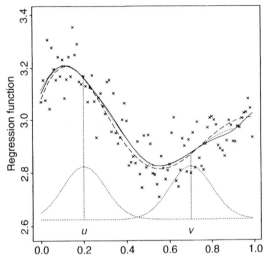

Figure 5.3. *Local cubic kernel estimate (solid curve) of the regression function m given by (5.1) based on* 100 *simulated observations (represented by crosses). The dashed curve is the true function m. The dotted curves are the kernel weights and cubic fits at the points u and v.*

Our next task is to derive an explicit expression for the local polynomial kernel estimator. We will do this for the fixed design case; the random design case is exactly the same except with X_i instead of x_i. Let p be the degree of the polynomial being fit. At a point x the estimator $\hat{m}(x; p, h)$ is obtained by fitting the polynomial

$$\beta_0 + \beta_1(\cdot - x) + \ldots + \beta_p(\cdot - x)^p$$

to the (x_i, Y_i) using weighted least squares with kernel weights $K_h(x_i - x)$. The value of $\hat{m}(x; p, h)$ is the height of the fit $\hat{\beta}_0$ where $\hat{\boldsymbol{\beta}} = (\hat{\beta}_0, \ldots, \hat{\beta}_p)^T$ minimises

$$\sum_{i=1}^{n} \{Y_i - \beta_0 - \ldots - \beta_p(x_i - x)^p\}^2 K_h(x_i - x).$$

Assuming the invertibility of $\mathbf{X}_x^T \mathbf{W}_x \mathbf{X}_x$, standard weighted least squares theory (Exercise 5.1) leads to the solution

$$\hat{\boldsymbol{\beta}} = (\mathbf{X}_x^T \mathbf{W}_x \mathbf{X}_x)^{-1} \mathbf{X}_x^T \mathbf{W}_x \mathbf{Y}$$

5.2 LOCAL POLYNOMIAL KERNEL ESTIMATORS

where $\mathbf{Y} = (Y_1, \ldots, Y_n)^T$ is the vector of responses,

$$\mathbf{X}_x = \begin{bmatrix} 1 & x_1 - x & \cdots & (x_1 - x)^p \\ \vdots & \vdots & \ddots & \vdots \\ 1 & x_n - x & \cdots & (x_n - x)^p \end{bmatrix}$$

is an $n \times (p+1)$ design matrix and

$$\mathbf{W}_x = \mathrm{diag}\{K_h(x_1 - x), \ldots, K_h(x_n - x)\}$$

is an $n \times n$ diagonal matrix of weights. Since the estimator of $m(x)$ is the intercept coefficient we obtain

$$\hat{m}(x; p, h) = \mathbf{e}_1^T (\mathbf{X}_x^T \mathbf{W}_x \mathbf{X}_x)^{-1} \mathbf{X}_x^T \mathbf{W}_x \mathbf{Y} \tag{5.2}$$

where $\mathbf{e}_1$ is the $(p+1) \times 1$ vector having 1 in the first entry and zero elsewhere.

Simple explicit formulae exist for the Nadaraya-Watson estimator ($p = 0$):

$$\hat{m}(x; 0, h) = \sum_{i=1}^n K_h(x_i - x) Y_i \Big/ \sum_{i=1}^n K_h(x_i - x) \tag{5.3}$$

and the local linear estimator ($p = 1$):

$$\hat{m}(x; 1, h) = n^{-1} \sum_{i=1}^n \frac{\{\hat{s}_2(x; h) - \hat{s}_1(x; h)(x_i - x)\} K_h(x_i - x) Y_i}{\hat{s}_2(x; h) \hat{s}_0(x; h) - \hat{s}_1(x; h)^2} \tag{5.4}$$

where

$$\hat{s}_r(x; h) = n^{-1} \sum_{i=1}^n (x_i - x)^r K_h(x_i - x)$$

(Exercise 5.2).

Because K is symmetric we could write $K_h(x - x_i)$ rather than $K_h(x_i - x)$, as for the kernel density estimator. We prefer the latter notation here because it emphasises the fact that the local polynomial kernel estimator is a weighted regression on the data, centred about each x.

5.3 Asymptotic MSE approximations: linear case

Because of their simplicity we will present the mean and variance calculations for $\hat{m}(x;p,h)$ in the case $p = 1$. It will also be assumed that x is a point in the "interior" of the design (see condition (iv) below). Extensions to general p and non-interior points will be postponed to Sections 5.4 and 5.5 respectively.

5.3.1 Fixed equally spaced design

Consider the fixed equally spaced design regression model

$$Y_i = m(x_i) + v^{1/2}(x_i)\varepsilon_i, \qquad i = 1,\ldots,n$$

where $x_i = i/n$. We will make the following assumptions in our analysis:

(i) The functions m'' and v are each continuous on $[0, 1]$.
(ii) The kernel K is symmetric about zero and is supported on $[-1, 1]$.
(iii) The bandwidth $h = h_n$ is a sequence satisfying $h \to 0$ and $nh \to \infty$ as $n \to \infty$.
(iv) The point x at which the estimation is taking place satisfies $h < x < 1 - h$ for all $n \geq n_0$ where n_0 is fixed.

In this subsection we will make the additional assumption that K has a bounded first derivative. Condition (iv) ensures that x is more than a bandwidth away from the boundary for all sufficiently large n.

It follows directly from (5.2) that

$$E\hat{m}(x;1,h) = \mathbf{e}_1^T(\mathbf{X}_x^T\mathbf{W}_x\mathbf{X}_x)^{-1}\mathbf{X}_x^T\mathbf{W}_x\mathbf{M}$$

where $\mathbf{M} = (m(x_1),\ldots,m(x_n))^T$. In the case of local linear fitting note that

$$\mathbf{X}_x = \begin{bmatrix} 1 & x_1 - x \\ \vdots & \vdots \\ 1 & x_n - x \end{bmatrix}.$$

According to a version of Taylor's theorem, for any $x \in [0, 1]$,

$$m(x_i) = m(x) + (x_i - x)m'(x) + \tfrac{1}{2}(x_i - x)^2 m''(x) + \ldots$$

5.3 ASYMPTOTIC MSE APPROXIMATIONS: LINEAR CASE

which implies that

$$\mathbf{M} = \mathbf{X}_x \begin{bmatrix} m(x) \\ m'(x) \end{bmatrix} + \tfrac{1}{2} m''(x) \begin{bmatrix} (x_1 - x)^2 \\ \vdots \\ (x_n - x)^2 \end{bmatrix} + \dots$$

The first term in the expansion of $E\hat{m}(x; 1, h)$ is therefore

$$\mathbf{e}_1^T (\mathbf{X}_x^T \mathbf{W}_x \mathbf{X}_x)^{-1} (\mathbf{X}_x^T \mathbf{W}_x \mathbf{X}_x) \begin{bmatrix} m(x) \\ m'(x) \end{bmatrix} = \mathbf{e}_1^T \begin{bmatrix} m(x) \\ m'(x) \end{bmatrix} = m(x),$$

the true regression function. This leads to the bias of $\hat{m}(x; 1, h)$ being

$$E\hat{m}(x; 1, h) - m(x)$$
$$= \tfrac{1}{2} m''(x) \, \mathbf{e}_1^T (\mathbf{X}_x^T \mathbf{W}_x \mathbf{X}_x)^{-1} \mathbf{X}_x^T \mathbf{W}_x \begin{bmatrix} (x_1 - x)^2 \\ \vdots \\ (x_n - x)^2 \end{bmatrix} + \dots \quad (5.5)$$

Notice that if m is linear then $m^{(r)}(x) = 0$ for all $r \geq 2$. Therefore $\hat{m}(x; 1, h)$ has the appealing property that it is exactly unbiased for linear m.

To compute the leading bias term for general m observe that

$$n^{-1} \mathbf{X}_x^T \mathbf{W}_x \mathbf{X}_x = \begin{bmatrix} \hat{s}_0(x; h) & \hat{s}_1(x; h) \\ \hat{s}_1(x; h) & \hat{s}_2(x; h) \end{bmatrix}$$

and

$$n^{-1} \mathbf{X}_x^T \mathbf{W}_x \begin{bmatrix} (x_1 - x)^2 \\ \vdots \\ (x_n - x)^2 \end{bmatrix} = \begin{bmatrix} \hat{s}_2(x; h) \\ \hat{s}_3(x; h) \end{bmatrix}.$$

Since K' is bounded, the entries of these matrices can be approximated by integrals as follows. From conditions (ii)-(iv), for n sufficiently large,

$$\hat{s}_\ell(x; h) = \int_0^1 (y - x)^\ell K_h(y - x) \, dy + O(n^{-1})$$
$$= h^\ell \int_{-x/h}^{(1-x)/h} u^\ell K(u) \, du + O(n^{-1})$$
$$= h^\ell \int_{-1}^{1} u^\ell K(u) \, du + O(n^{-1}).$$

By the symmetry and compact support of K the odd moments of K vanish so we are left with

$$n^{-1}\mathbf{X}_x^T\mathbf{W}_x\mathbf{X}_x = \begin{bmatrix} 1+O(n^{-1}) & O(n^{-1}) \\ O(n^{-1}) & h^2\mu_2(K)+O(n^{-1}) \end{bmatrix}$$

and

$$\mathbf{X}_x^T\mathbf{W}_x \begin{bmatrix} (x_1-x)^2 \\ \vdots \\ (x_n-x)^2 \end{bmatrix} = \begin{bmatrix} h^2\mu_2(K)+O(n^{-1}) \\ O(n^{-1}) \end{bmatrix}$$

where $\mu_\ell(K) = \int_{-1}^{1} z^\ell K(z)\,dz$, as before. Some simple matrix algebra leads to the expression

$$E\hat{m}(x;1,h) - m(x) = \tfrac{1}{2}h^2 m''(x)\mu_2(K) + o(h^2) + O(n^{-1})$$

for the leading bias term.

For the variance approximation note that

$$\mathrm{Var}\{\hat{m}(x;1,h)\}$$
$$= \mathbf{e}_1^T(\mathbf{X}_x^T\mathbf{W}_x\mathbf{X}_x)^{-1}\mathbf{X}_x^T\mathbf{W}_x\mathbf{V}\mathbf{W}_x\mathbf{X}_x(\mathbf{X}_x^T\mathbf{W}_x\mathbf{X}_x)^{-1}\mathbf{e}_1$$

where

$$\mathbf{V} = \mathrm{diag}\{v(x_1),\ldots,v(x_n)\}.$$

Using approximations analogous to those used above,

$$n^{-1}\mathbf{X}_x^T\mathbf{W}_x\mathbf{V}\mathbf{W}_x\mathbf{X}_x$$
$$= n^{-1}\sum_{i=1}^{n} K_h(x_i-x)^2 v(x_i) \begin{bmatrix} 1 & x_i-x \\ x_i-x & (x_i-x)^2 \end{bmatrix}$$
$$= \begin{bmatrix} h^{-1}R(K)v(x)+o(h^{-1}) & O(n^{-1}) \\ O(n^{-1}) & h\mu_2(K^2)v(x)+O(n^{-1}) \end{bmatrix}$$

where, again, $R(K) = \int K(z)^2\,dz$. These can be combined to obtain

$$\mathrm{Var}\{\hat{m}(x;1,h)\} = (nh)^{-1}R(K)v(x) + o\{(nh)^{-1}\}.$$

5.3.2 Random design

Now suppose that the design is an independent sample, denoted by $X_1, \ldots, X_n$, having density f. For simplicity's sake we will assume here that f has support on $[0,1]$ and that conditions (i)–(iv) from the previous section are satisfied. In addition it is assumed that f' is continuous.

In the random design setting the bias and variance calculations can be done analogously to the fixed design case provided we condition on the predictor variables. Technical problems arise if one integrates over the predictor variables because the denominator of $\hat{m}(x; 1, h)$ can equal zero with positive probability. We prefer to avoid the necessary adjustments and work with conditional bias and variance. It is sometimes argued that one should not average over design variables in any case.

If we replace $\mathbf{X}_x$ and $\mathbf{W}_x$ used in Section 5.2 by

$$\mathbf{X}_x = \begin{bmatrix} 1 & X_1 - x & \cdots & (X_1 - x)^p \\ \vdots & \vdots & \ddots & \vdots \\ 1 & X_n - x & \cdots & (X_n - x)^p \end{bmatrix}$$

and

$$\mathbf{W}_x = \mathrm{diag}\{K_h(X_i - x), \ldots, K_h(X_n - x)\}$$

then the arguments used to derive (5.5) lead to

$$E\{\hat{m}(x; 1, h) - m(x) | X_1, \ldots, X_n\}$$

$$= \tfrac{1}{2} m''(x) \mathbf{e}_1^T (\mathbf{X}_x^T \mathbf{W}_x \mathbf{X}_x)^{-1} \mathbf{X}_x^T \mathbf{W}_x \begin{bmatrix} (X_1 - x)^2 \\ \vdots \\ (X_n - x)^2 \end{bmatrix} + \cdots.$$

Once again we see that if $m(x)$ is linear then $\hat{m}(x; 1, h)$ is conditionally unbiased given $X_1, \ldots, X_n$. Let o_P denote "smaller order in probability" (see Appendix A). To approximate the leading conditional bias of $\hat{m}(x; 1, h)$ analyses similar to those used in Chapter 2 for the kernel density estimator give

$$\hat{s}_\ell(x; h) = \begin{cases} h^\ell \mu_\ell(K) f(x) + o_P(h^\ell) & \ell \text{ even} \\ h^{\ell+1} \mu_{\ell+1}(K) f'(x) + o_P(h^{\ell+1}) & \ell \text{ odd} \end{cases} \quad (5.6)$$

which leads to

$$n^{-1} \mathbf{X}_x^T \mathbf{W}_x \mathbf{X}_x$$
$$= \begin{bmatrix} f(x) + o_P(1) & h^2 f'(x) \mu_2(K) + o_P(h^2) \\ h^2 f'(x) \mu_2(K) + o_P(h^2) & h^2 f(x) \mu_2(K) + o_P(h^2) \end{bmatrix} \quad (5.7)$$

and

$$n^{-1}\mathbf{X}_x^T\mathbf{W}_x \begin{bmatrix} (X_1 - x)^2 \\ \vdots \\ (X_n - x)^2 \end{bmatrix} = \begin{bmatrix} h^2 f(x)\mu_2(K) + o_P(h^2) \\ h^4 f'(x)\mu_4(K) + o_P(h^4) \end{bmatrix}. \quad (5.8)$$

Notice that

$$(n^{-1}\mathbf{X}_x^T\mathbf{W}_x\mathbf{X}_x)^{-1}$$
$$= \begin{bmatrix} f(x)^{-1} + o_P(1) & -f'(x)/f(x)^2 + o_P(1) \\ -f'(x)/f(x)^2 + o_P(1) & \{h^2 f(x)\mu_2(K)\}^{-1} + o_P(h^{-2}) \end{bmatrix}.$$

It follows that the conditional bias is given by

$$\begin{aligned} E\{\hat{m}(x;1,h) - m(x)|X_1,\ldots,X_n\} \\ = \tfrac{1}{2}h^2 m''(x)\mu_2(K) + o_P(h^2). \end{aligned} \quad (5.9)$$

Analogous manipulations for the variance show that

$$\begin{aligned} \text{Var}\{\hat{m}(x;1,h)|X_1,\ldots,X_n\} \\ = \{(nh)^{-1}R(K)/f(x)\}v(x) + o_P\{(nh)^{-1}\} \end{aligned} \quad (5.10)$$

(Fan, 1992a).

These expressions have an intuitively simple interpretation. The leading bias term depends on x only through $m''(x)$ which reflects the error of the linear approximation. If m is close to being linear at x then $m''(x)$ is relatively small, which is consistent with the notion of local linear fits having less bias in this case. On the other hand, if m has a high amount of curvature at x then $m''(x)$ is higher and local linear fits tend to produce more biased estimates. The fact that the bias depends on h^2 also reflects the notion that bias is increased with more smoothing. Indeed, expression (5.9) is a direct analogue of the bias approximation for kernel density estimation given by (2.10).

If K is the density of the uniform distribution, so that all non-zero weights are constant, then the expression in curly brackets in (5.10) is approximately the reciprocal of the sample size used in the local fit. Otherwise, this expression can be thought of as the reciprocal of the "effective local sample size". Thus, (5.10) reflects the fact that the variance will be penalised by larger conditional variance and sparser data near x. Note that (5.10) reduces to the variance of the equally spaced design case when f is uniform.

5.4 Asymptotic MSE approximations: general case

We now take a brief look at the extension of the MSE approximations of the previous section to the case of general polynomial degree. Emphasis will be on the MSE results themselves rather than their derivation since the requisite manipulations are somewhat more involved (see Ruppert and Wand, 1994). In this section we assume that the conditions (i)–(iv) are satisfied, with the additional assumption that $m^{(p+2)}(x)$ is continuous.

Useful notation for concise presentation of results is as follows. Let $\mathbf{N}_p$ be the $(p+1) \times (p+1)$ matrix having (i,j) entry equal to $\mu_{i+j-2}(K)$ and let $\mathbf{M}_p(u)$ be the same as $\mathbf{N}_p$, but with the first column replaced by $(1, u, ..., u^p)^T$. Then the kernel

$$K_{(p)}(u) = \{|\mathbf{M}_p(u)|/|\mathbf{N}_p|\}K(u) \tag{5.11}$$

is a $(p+1)$th-order kernel when p is odd, and a $(p+2)$th-order when p is even. In fact, it can be shown that $K_{(p)} = K_{(p+1)}$ if p is even. For example,

$$K_{(2)}(u) = K_{(3)}(u) = \frac{\mu_4(K) - \mu_2(K)u^2}{\mu_4(K) - \mu_2(K)^2}K(u)$$

(this is a general formula for constructing a fourth-order kernel from a second-order kernel; see Section 2.8). For odd p the conditional bias of $\hat{m}(x; p, h)$ is

$$E\{\hat{m}(x; p, h) - m(x)|X_1, \ldots, X_n\}$$
$$= \frac{1}{(p+1)!}h^{p+1}m^{(p+1)}(x)\mu_{p+1}(K_{(p)}) + o_P(h^{p+1})$$

while for even p

$$E\{\hat{m}(x; p, h) - m(x)|X_1, \ldots, X_n\}$$
$$= h^{p+2}\Big\{\frac{1}{(p+1)!}m^{(p+1)}(x)f'(x)/f(x) \tag{5.12}$$
$$+ \frac{1}{(p+2)!}m^{(p+2)}(x)\Big\}\mu_{p+2}(K_{(p)}) + o_P(h^{p+2}).$$

In either case

$$\mathrm{Var}\{\hat{m}(x; p, h)|X_1, \ldots, X_n\}$$
$$= \{(nh)^{-1}R(K_{(p)})/f(x)\}v(x) + o_P\{(nh)^{-1}\}. \tag{5.13}$$

The first point to note is that the degree of polynomial being locally fitted determines the order of the bias of $\hat{m}(x;p,h)$. Indeed, in terms of kernel dependent constants, fitting a pth degree polynomial using kernel K is similar to what one would expect if using the $2(\lfloor p/2 \rfloor +1)$th-order kernel $K_{(p)}$ given by (5.11) in density estimation. (Here $\lfloor x \rfloor$ denotes the greatest integer not exceeding x.) Thus, if K is a second-order kernel then local constant and linear fits correspond to second-order kernel estimation, local quadratic and cubic fits are like fourth-order kernel estimation, and so on. However, it is only odd degree polynomial fits that have approximate bias depending x only through $m^{(p+1)}(x)$. The simplicity and interpretability of such a bias term is appealing since the $(p+1)$th derivative of m reflects the error due to fitting pth degree polynomials, as well as being directly analogous to the behaviour of kernel density estimation using the $(p+1)$th-order kernel $K_{(p)}$. On the other hand, even degree polynomial kernel estimators have a more complicated bias expression which does not lend itself to simple interpretation. This extra complexity also makes extension of plug-in bandwidth selection rules more difficult. The relative performance of consecutive odd and even degree polynomial fits in MSE terms is not clear-cut, as it depends on the relative values of m and f and their derivatives. In the special case of uniformly distributed design variables the distinction between asymptotic bias of pairs of consecutive even and odd degree fits disappears since $f'(x) = 0$ for all x.

An important problem is the choice of p. The results in this section show that, for sufficiently smooth regression functions, the asymptotic performance of $\hat{m}(\cdot;p,h)$ improves for higher values of p. However, the *practical* gains from higher-degree fits are not as clear-cut. As with higher-order kernels, the variance of the estimator becomes larger for higher p and, in many situations, a very large sample may be required for there to be a substantial improvement in practical performance, especially beyond cubic fits. Odd degree fits have attractive bias and boundary properties (see Section 5.5). These facts suggests use of either $p = 1$ or $p = 3$.

5.5 Behaviour near the boundary

The analyses of the previous two sections were performed under the assumption that x is an "interior point" of the support of $[0,1]$. We now investigate the performance of local polynomial estimators near the boundary. The essential reason for boundary

5.5 BEHAVIOUR NEAR THE BOUNDARY

performance requiring special treatment is the fact that part of the kernel window is devoid of data. This is shown in Figure 5.4 for the Epanechnikov kernel with bandwidth $h = 0.20$, centred about $u = 0.08$, $v = 0.45$ and $w = 0.90$. In this case u is a left boundary point, since its associated kernel weight over-spills the left boundary. For similar reasons, w is a right boundary point, while v is an interior point.

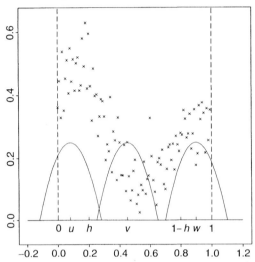

Figure 5.4. *Illustration of a left boundary point (u), an interior point (v) and a right boundary point (w) for a design density supported on* $[0, 1]$. *The kernel is the Epanechnikov with bandwidth* $h = 0.2$.

Suppose that $x = x_n = \alpha h$ where $0 \le \alpha < 1$; that is, x is a sequence of points that is within αh of the left boundary of $[0, 1]$ for all $h > 0$. Treatment of the right boundary is analogous. This supposition was also made in Section 2.11 in the density estimation context, and allows a useful quantification of behaviour of $\hat{m}(x; p, h)$ near the boundary. Incomplete moments of K are denoted by

$$\mu_{\ell,\alpha}(K) = \int_{-\alpha}^{1} z^{\ell} K(z) \, dz.$$

Note that this definition of incomplete moment differs slightly from that given in Section 2.11. This is because of the differences in what have become the conventional definitions of the kernel density estimator and the local polynomial kernel estimator concerning the argument of the function K_h. Reworking the arguments of Section

5.3 we see that (5.7) for boundary points becomes

$$n^{-1}\mathbf{X}_x^T\mathbf{W}_x\mathbf{X}_x = f(x)\begin{bmatrix} \mu_{0,\alpha}(K) + o_P(1) & h\mu_{1,\alpha}(K) + o_P(h) \\ h\mu_{1,\alpha}(K) + o_P(h) & h^2\mu_{2,\alpha}(K) + o_P(h^2) \end{bmatrix}$$

and (5.8) becomes

$$n^{-1}\mathbf{X}_x^T\mathbf{W}_x\begin{bmatrix} (X_1 - x)^2 \\ \vdots \\ (X_n - x)^2 \end{bmatrix} = f(x)\begin{bmatrix} h^2\mu_{2,\alpha}(K) + o_P(h^2) \\ h^3\mu_{3,\alpha}(K) + o_P(h^3) \end{bmatrix}.$$

This leads to a bias expression of the form

$$\begin{aligned} E\{\hat{m}(x;1,h) &- m(x)|X_1,\ldots,X_n\} \\ &= \tfrac{1}{2}h^2 m''(x)Q_\alpha(K) + o_P(h^2) \end{aligned} \tag{5.14}$$

where

$$Q_\alpha(K) = \frac{\mu_{2,\alpha}^2(K) - \mu_{1,\alpha}(K)\mu_{3,\alpha}(K)}{\mu_{2,\alpha}(K)\mu_{0,\alpha}(K) - \mu_{1,\alpha}^2(K)}.$$

Therefore, the boundary bias of $\hat{m}(x;1,h)$ is of the same form as the interior, except that the kernel dependent constant is different.

Expressions of this type hold for general local polynomial kernel estimators. One can give concise results for x near the boundary by extending the $K_{(p)}$ notation of the previous section to truncated moments. Define

$$K_{(p)}(u,\alpha) = \{|M_p(u,\alpha)|/|N_p(\alpha)|\}K(u), \quad -\alpha < u < 1 \tag{5.15}$$

where $N_p(\alpha)$ is the $(p+1) \times (p+1)$ matrix having (i,j) entry equal to $\mu_{i+j-2,\alpha}(K)$ and $M_p(u,\alpha)$ is the same as $N_p(\alpha)$, but with the first column replaced by $(1,u,\ldots,u^p)^T$. Then it can be shown that the left-hand boundary bias of $\hat{m}(x;p,h)$ is

$$\begin{aligned} &E\{\hat{m}(x;p,h) - m(x)|X_1,\ldots,X_n\} \\ &= \frac{1}{(p+1)!}h^{p+1}\left\{\int_{-\alpha}^1 u^{p+1}K_{(p)}(u,\alpha)\,du\right\}m^{(p+1)}(x) + o_P(h^{p+1}). \end{aligned}$$

Similarly, it can be shown that the conditional variance of $\hat{m}(x;p,h)$ is asymptotic to

$$\begin{aligned} &\operatorname{Var}\{\hat{m}(x;p,h)|X_1,\ldots,X_n\} \\ &= (nh)^{-1}\left\{\int_{-\alpha}^1 K_{(p)}(u,\alpha)^2\,du\right\}v(x)/f(x) + o_P\{(nh)^{-1}\}. \end{aligned}$$

5.5 BEHAVIOUR NEAR THE BOUNDARY

In a sense these expressions are a generalisation of the results presented in Section 5.4 since those results can be obtained by setting $\alpha \geq 1$. However, note that for even p

$$\int_{-1}^{1} u^{p+1} K_{(p)}(u, 1)\, du = 0$$

which leads to the more complicated $O(h^{p+2})$ leading bias expressions given by (5.12).

A discrepancy between the orders of magnitude of bias in the interior and near the boundary is usually referred to as a *boundary bias* problem. Suppose that p is even. In the interior the usual variance-bias trade-off leads to an optimal bandwidth of order $n^{-1/(2p+5)}$ and corresponding optimal minimum MSE of order $n^{-(2p+4)/(2p+5)}$. However, near the boundary the optimal bandwidth is of order $n^{-1/(2p+3)}$ and the minimum MSE is of order $n^{-(2p+2)/(2p+3)}$. For the Nadaraya-Watson estimator ($p = 0$) this means that the usual $n^{-4/5}$ optimal MSE is inflated to $n^{-2/3}$ near the boundaries. While this is merely an asymptotic argument, boundary effects are also quite noticeable in practice; for example see Figure 5.7 in the next section. As discussed in Section 2.11, this has motivated the use of suitably modified kernels near the boundary that correct this asymptotic discrepancy. Further discussion of boundary kernels in the regression context is given in the next section.

For p odd $\int_{-1}^{1} u^{p+1} K_{(p)}(u, 1)\, du$ is typically non-zero and so the bias of $\hat{m}(x; p, h)$ is proportional to $h^{p+1} m^{(p+1)}(x)$ both in the interior and near the boundaries. Therefore, in terms of rates of convergence of the bias, no boundary adjustment is necessary. This is a great advantage of local linear fitting over the Nadaraya-Watson estimator and other traditional kernel estimators. However, the asymptotic variance tends to become inflated near the boundary. Figure 5.5 shows a plot of

$$\int_{-\alpha}^{1} K_{(1)}(u, \alpha)\, du \Big/ \int_{-\alpha}^{1} K_{(0)}(u, \alpha)\, du$$

$$\simeq \operatorname{Var}\{\hat{m}(x; 1, h)\} / \operatorname{Var}\{\hat{m}(x; 0, h)\}$$

for $0 \leq \alpha \leq 1$ and the biweight kernel $K(u) = \frac{15}{16}(1 - u^2)^2 1_{\{|u|<1\}}$, under the assumption that the same bandwidth is used for each. In particular, the variance of $\hat{m}(x; 1, h)$ is about 3.58 times that of $\hat{m}(x; 0, h)$ at the boundary itself.

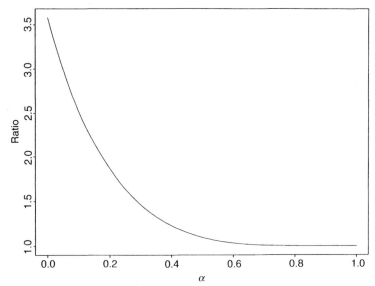

Figure 5.5. *Ratio of asymptotic boundary variances of the local linear kernel estimator to that of the Nadaraya-Watson estimator for $0 \leq \alpha \leq 1$, based on the biweight kernel.*

5.6 Comparison with other kernel estimators

5.6.1 *Asymptotic comparison*

The favourable properties of local polynomial kernel estimators, especially those of odd degree, are a relatively recent discovery (Fan, 1992a, 1993, Hastie and Loader, 1993). In its earlier years, the kernel regression literature was essentially confined to the study and use of the Nadaraya-Watson estimator

$$\hat{m}_{\text{NW}}(x;h) \equiv \hat{m}(x;0,h) = \sum_{i=1}^{n} K_h(X_i - x)Y_i \bigg/ \sum_{i=1}^{n} K_h(X_i - x)$$

(Nadaraya, 1964, Watson, 1964), the *Priestley-Chao* estimator

$$\hat{m}_{\text{PC}}(x;h) = \sum_{i=1}^{n} \{X_{(i)} - X_{(i-1)}\} K_h(X_{(i)} - x) Y_{[i]}$$

5.6 COMPARISON WITH OTHER KERNEL ESTIMATORS

(Priestley and Chao, 1972) and the *Gasser-Müller* estimator

$$\hat{m}_{\text{GM}}(x;h) = \sum_{i=1}^{n} \int_{s_{i-1}}^{s_i} K_h(u-x)\,du\, Y_{[i]}$$

where $s_i = \frac{1}{2}(X_{(i)} + X_{(i+1)})$, $s_0 = 0$, $s_n = 1$ (Gasser and Müller, 1979). Here $(X_{(i)}, Y_{[i]})$, $i = 1, \ldots, n$, denote the (X_i, Y_i) ordered with respect to the X_i values.

As we mentioned in Section 5.1, it is easy to see that the Nadaraya-Watson estimator can be obtained from local constant fitting, that is $\hat{m}_{\text{NW}}(\cdot;h) = \hat{m}(\cdot;0,h)$, although this was not the original motivation for the method; see Exercise 5.5. From the results of the previous two sections it follows that the bias and variance of $\hat{m}_{\text{NW}}(\cdot;h)$ and $\hat{m}(\cdot;1,h)$ are each of the same order of magnitude in the interior of the support of f. However, $\hat{m}_{\text{NW}}(\cdot;h)$ has a more complicated bias expression and $O(h)$ boundary bias.

The simple bias of $\hat{m}(\cdot;1,h)$ is shared by the Priestley-Chao and Gasser-Müller estimators but both of these estimators have boundary bias problems. Moreover, for random designs,

$$\text{Var}\{\hat{m}_{\text{GM}}(x;h)\} \sim \tfrac{3}{2}[\text{Var}\{\hat{m}(x;1,h)\}] \qquad (5.16)$$

(Chu and Marron, 1991). A similar result holds for the Priestley-Chao estimator (e.g. Mack and Müller, 1989). This 50% increase in variance is due to the inefficient use of the data by estimators of this type (see Chu and Marron, 1991). Therefore, with respect to MSE, $\hat{m}_{\text{GM}}(x;h)$ and $\hat{m}_{\text{PC}}(x;h)$ are asymptotically inadmissible as estimators of $m(x)$.

The following discussion provides some insight into the more appealing properties of $\hat{m}(x;1,h)$ compared to those of $\hat{m}_{\text{NW}}(x;h)$ with regard to their respective adjustments for non-uniform designs. It is straightforward to show that the kernel weighted linear combination of the Y_i's

$$\hat{a}(x;h) = n^{-1} \sum_{i=1}^{n} K_h(X_i - x) Y_i$$

is a consistent estimator for $(mf)(x)$. If f is known then two consistent estimators for $m(x)$ are

$$\hat{m}_1(x;h) = n^{-1} f(x)^{-1} \sum_{i=1}^{n} K_h(X_i - x) Y_i \qquad (5.17)$$

and

$$\hat{m}_2(x;h) = n^{-1} \sum_{i=1}^{n} f(X_i)^{-1} K_h(X_i - x) Y_i. \quad (5.18)$$

The first estimator is, perhaps, more obvious since it simply divides a consistent estimator for $(mf)(x)$ by $f(x)$. However, it is $\hat{m}_2(x;h)$ that has the more attractive bias (see Exercise 5.9), as well as being more natural in other ways – see the discussion in the last paragraph of Section 6.2.2 for more on this naturalness.

The Nadaraya-Watson estimator takes the design density into account by dividing $\hat{a}(x;h)$ by the kernel density estimator $\hat{f}(x;h) \simeq f(x)$ which leads to an estimate with bias properties similar to $\hat{m}_1(x;h)$. On the other hand, the weight that multiplies $K_h(X_i - x)Y_i$ in $\hat{m}(x;1,h)$ is

$$\frac{\hat{s}_2(x;h) - \hat{s}_1(x;h)(X_i - x)}{\hat{s}_2(x;h)\hat{s}_0(x;h) - \hat{s}_1(x;h)^2}$$

which, using the approximations in (5.6), is asymptotic to

$$(1/f)(x) + (X_i - x)(1/f)'(x) \simeq (1/f)(X_i).$$

This leads to $\hat{m}(x;1,h)$ having bias properties similar to those of $\hat{m}_2(x;h)$.

Equivalence results have been established between local polynomial kernel estimators and kernel estimators of the Priestley-Chao or Gasser-Müller type in the case of fixed designs (Lejeune, 1985, Müller, 1987). These results essentially state that the quotients of the weights applied to the Y_i's by each type of estimator tend to unity as $n \to \infty$. For random designs (5.16) above indicates that this equivalence no longer holds. However, one can adjust Priestley-Chao estimators to overcome their inefficiency. One such adjustment is

$$\tilde{m}_{\text{PC}}(x;h,s) = \sum_{i=1}^{n} \tfrac{1}{2s}(X_{(i-s)} - X_{(i+s)}) K_h(X_{(i)} - x) Y_{[i]} \quad (5.19)$$

where $s/n \to \infty$ (Jones, Davies and Park, 1994) which has the same asymptotic bias and variance as $\hat{m}(x;1,h)$, suggesting an asymptotic equivalence between these two estimators. Furthermore, if p is odd and the $(p+1)$th-order kernel $K_{(p)}$ is used instead of the second-order kernel K in (5.19) then the main conditional MSE

5.6 COMPARISON WITH OTHER KERNEL ESTIMATORS

terms of the resulting estimator are the same as those of $\hat{m}(x;p,h)$. This suggests an equivalence between $(p+1)$th-order versions of $\widetilde{m}_{PC}(x;h,s)$ and local pth degree kernel estimators for p odd.

Conditional MSE results near the boundary suggest that the behaviour of local kernel estimators of odd degree is equivalent to that of other kernel estimators with boundary kernels used near the boundary. If a kth-order kernel with support on $[-1,1]$ is used in the interior then the boundary kernel for use within αh of the boundary is a kernel $B_{[k]}(\cdot,\alpha)$ on $[-\alpha,1]$ that satisfies

$$\int_{-\alpha}^{1} u^j B_{[k]}(u,\alpha)\,du = \begin{cases} 1 & j=0 \\ 0 & j=1,\ldots,k-1 \\ \beta_\alpha \neq 0 & j=k \end{cases}$$

for each $0 \le \alpha \le 1$. These properties ensure that the order of the bias near the boundary is the same as in the interior. However, for p odd, the kernel $K_{(p)}(\cdot,\alpha)$ defined by (5.15) can also be shown to satisfy the above properties of $B_{[k]}(\cdot,\alpha)$ when $k=p+1$, and is therefore a boundary kernel of order $p+1$. Formula (5.15) is one of a variety of possible formulae for obtaining a boundary kernel from a given "interior" one. Special cases of (5.15) have appeared in various places in the literature (e.g. Gasser and Müller, 1979), and can be shown to be as good as any of the others in terms of performance (Jones, 1993). The $p=1$ version is precisely K^L as discussed in Section 2.11. It is easily shown that, for p odd, the conditional MSE expressions of $\hat{m}(x;p+1,h)$ near the boundary match those of $\widetilde{m}_{PC}(x;h,s)$ with $K_{(p+1)}(\cdot,\alpha)$ used near the left boundary. Therefore, there is a sense in which odd degree local polynomial kernel estimators automatically induce a boundary kernel-type bias correction.

5.6.2 Effective kernels

A very informative graphical tool for comparing different types of kernel regression estimators (and other nonparametric regression estimators) is the *effective kernel* of the estimator (e.g. Hastie and Loader, 1993). Observe that all kernel-type estimators of $m(x)$ can be written in the form

$$\hat{m}(x) = \sum_{i=1}^{n} w_i(x) Y_i.$$

The *effective kernel* at x is defined to be the set of weights $w_i(x)$, $i=1,\ldots,n$. Important insight into the properties of a particular estimator can be obtained by plotting the effective kernel against the X_i's.

Figure 5.6 displays effective kernel plots for estimation of $m(0.5)$, where $m(x) = \sin(2\pi x)$ on $[0,1]$, using

(a) $\hat{m}_{\text{NW}}(0.5; 0.2)$, (b) $\hat{m}_{\text{GM}}(0.5; 0.2)$ and (c) $\hat{m}(0.5; 1, 0.2)$.

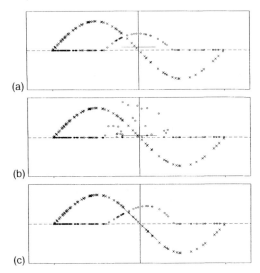

Figure 5.6. *Effective kernels, indicated by the circles, for estimation of $m(0.5)$ using (a) the Nadaraya-Watson estimator, (b) the Gasser-Müller estimator and (c) the local linear estimator. The short solid lines correspond to the estimates of $m(0.5)$ and the crosses indicate the data.*

The kernel is the Epanechnikov kernel on $[-1, 1]$. For clarity of presentation the data are observed without error and the effective kernel weights have been multiplied by a factor of 12. In this case the asymmetry of the observations combined with the slope of the regression function leads to an upwardly biased Nadaraya-Watson estimate. The effective kernel of $\hat{m}_{\text{NW}}(x; h)$ is symmetric and therefore unable to adjust for the asymmetry. The Gasser-Müller estimate does allow adjustment, but notice how noisy the weights are, even though neighbouring observations contain similar information about $m(0.5)$. This is related to the inefficiency of such estimators: if the data were observed with error the estimate would be quite variable. The local linear estimate, however, is able to adapt to the design and overcome the inefficiency in a much more appealing manner. Notice the effective kernel where the observations on the left are downweighted, while those on the right are upweighted, leading to an estimate with virtually no bias at $x = 0.5$.

The design adaptive qualities of $\hat{m}(x; 1, h)$ are even more apparent for estimation of the boundary value $m(0)$. This is demonstrated via effective kernel plots in Figure 5.7.

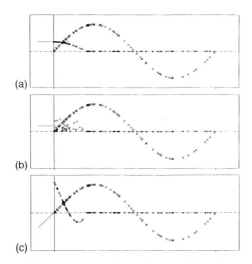

Figure 5.7. *Effective kernels, indicated by the circles, for estimation of $m(0)$ using (a) the Nadaraya-Watson estimator, (b) the Gasser-Müller estimator and (c) the local linear estimator. The short solid lines correspond to the estimates of $m(0)$ and the crosses indicate the data.*

In (a) and (b) the boundary bias problems of the Nadaraya-Watson and Gasser-Müller estimators are very apparent. However, the effective kernel of the local linear estimate shown in (c) deforms near the boundary to the extent of assigning negative weight to those observations farther away from the boundary, which drastically reduces the bias.

5.7 Derivative estimation

There are many situations in which derivatives of the regression function are the main focus of an investigation. For example, in the study of human growth curves, the first two derivatives of height as a function of age, representing "speed" and "acceleration" of growth respectively, have important biological significance (Müller, 1988). Estimation of derivatives of m is also required for plug-in bandwidth selection strategies, as described in the next section.

The extension of local polynomial fitting ideas to estimation of the rth derivative is straightforward. One can estimate $m^{(r)}(x)$ via the intercept coefficient of the rth derivative of the local polynomial being fitted at x, assuming that $r \leq p$. For example, the local polynomial estimate of $m'(x)$ is simply the slope of the local polynomial fit. This is illustrated in Figure 5.8 for a fictitious data set.

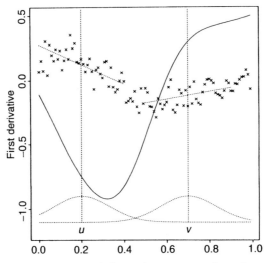

Figure 5.8. *Estimate of first derivative of m based on a local linear fit. The dotted lines show the locally fitted lines at two points u and v. The estimate of m' is the slope of the line at that point.*

In this case the estimate of m' at two points u and v is found from the slope of the local linear estimator. When this is performed at all points the solid curve results.

In general, the local pth degree estimate of $m^{(r)}(x)$ is

$$\widehat{m^{(r)}}(x;p,h) = r!\mathbf{e}_{r+1}^T(\mathbf{X}_x^T\mathbf{W}_x\mathbf{X}_x)^{-1}\mathbf{X}_x\mathbf{W}_x\mathbf{Y}$$

for all $r = 0, \ldots, p$. Here $\mathbf{e}_{r+1}$ is the $(p+1) \times 1$ vector having 1 in the $(r+1)$th entry and zeros elsewhere. Note that $\widehat{m^{(r)}}(x;p,h)$ is not, in general, equal to the rth derivative of $\hat{m}(x;p,h)$.

Bias and variance approximations again provide important insight into the performance of $\widehat{m^{(r)}}(x;p,h)$. This time it is conve-

5.7 DERIVATIVE ESTIMATION

nient to define

$$K_{(r,p)}(u) = r!\{|\mathbf{M}_{r,p}(u)|/|\mathbf{N}_p|\}K(u)$$

where $\mathbf{N}_p$ is as defined in Section 5.4, and $\mathbf{M}_{r,p}(u)$ equals $\mathbf{N}_p$ except that the $(r+1)$th column is replaced by $(1, u, ..., u^p)^T$. Let x be an interior point and suppose that $m^{(p+2)}(x)$ is continuous. Then for $p - r$ odd,

$$E\{\widehat{m^{(r)}}(x; p, h) - m^{(r)}(x)|X_1, \ldots, X_n\}$$
$$= \frac{1}{(p+1)!}\mu_{p+1}(K_{(r,p)})m^{(p+1)}(x)h^{p-r+1} + o_P(h^{p-r+1})$$

while for $p - r$ even,

$$E\{\widehat{m^{(r)}}(x; p, h) - m^{(r)}(x)|X_1, \ldots, X_n\}$$
$$= \Big[\frac{1}{(p+2)!}\mu_{p+2}(K_{(r,p)})m^{(p+2)}(x) + \frac{1}{(p+1)!}\{\mu_{p+2}(K_{(r,p)})$$
$$- r\mu_{p+1}(K_{(r-1,p)})\}m^{(p+1)}(x)f'(x)/f(x)\Big]h^{p-r+2}$$
$$+ o_P(h^{p-r+2}).$$

In either case

$$\mathrm{Var}\{\widehat{m^{(r)}}(x; p, h)\}$$
$$= n^{-1}h^{-2r-1}R(K_{(r,p)})v(x)/f(x) + o_P(n^{-1}h^{-2r-1})$$

(Ruppert and Wand, 1994). Therefore, $\widehat{m^{(r)}}(x; p, h)$ exhibits the simpler bias expression when $p - r$ is odd. For example, for estimation of $m'(x)$ the usual second-order kernel bias rate of order h^2 will be achieved by the slope of either local linear or local quadratic fitting. However, it is the slope of local quadratics that has the desirable property of the leading bias term depending only on $m'''(x)$, and not on $m''(x)f'(x)/f(x)$ as well. Likewise, for estimation of $m''(x)$, which is required for plug-in bandwidth selection (see Section 5.8), the lowest degree estimator having the less complicated bias expression is the second derivative associated with local cubic fitting. It can also be shown that the boundary bias of $\widehat{m^{(r)}}(x; p, h)$ is the same order as the interior bias when $p - r$ is odd.

5.8 Bandwidth selection

Each of the bandwidth selection ideas described in Chapter 3 for kernel density estimation can be adapted to kernel regression contexts. In this section we will briefly describe a version of the simple direct plug-in idea for local linear regression that has been shown to possess attractive theoretical and practical properties (Ruppert, Sheather and Wand, 1995).

For simplicity, assume that the errors are homoscedastic with common variance σ^2 and that the X_i's are from a compactly supported density on $[0, 1]$. An appropriate global error criterion is the weighted conditional MISE

$$\text{MISE}\{\hat{m}(\cdot; 1, h) | X_1, \ldots, X_n\}$$
$$= E\left[\int \{\hat{m}(x; 1, h) - m(x)\}^2 f(x)\, dx \Big| X_1, \ldots, X_n\right].$$

This weighting by f puts more emphasis on those regions where there are more data as well as simplifying the plug-in methodology. With respect to this criterion the asymptotically optimal bandwidth is

$$h_{\text{AMISE}} = C_1(K) \left[\frac{\sigma^2}{\theta_{22} n}\right]^{1/5}$$

where $C_1(K) = \{R(K)/\mu_2(K)^2\}^{1/5}$ and θ_{22} is a special case of the notation

$$\theta_{rs} = \int m^{(r)}(x) m^{(s)}(x) f(x)\, dx.$$

A natural estimator for θ_{22} is

$$\hat{\theta}_{22}(g) = n^{-1} \sum_{i=1}^{n} \widehat{m^{(2)}}(X_i; 3, g)^2$$

while a natural estimator for σ^2 is

$$\hat{\sigma}^2(\lambda) = \nu^{-1} \sum_{i=1}^{n} \{Y_i - \hat{m}(X_i; 1, \lambda)\}^2 \tag{5.20}$$

where

$$\nu = n - 2 \sum_{i=1}^{n} w_{ii} + \sum_{i=1}^{n} \sum_{j=1}^{n} w_{ij}^2$$

5.8 BANDWIDTH SELECTION

and
$$w_{ij} = \mathbf{e}_1^T (\mathbf{X}_{X_i}^T \mathbf{W}_{X_i} \mathbf{X}_{X_i})^{-1} \mathbf{X}_{X_i}^T \mathbf{W}_{X_i} \mathbf{e}_j.$$

Here $\mathbf{W}_{X_i}$ is based on the bandwidth λ. The naturalness of $\hat{\sigma}^2(\lambda)$ arises from the fact that it is exactly conditionally unbiased for σ^2 whenever m is linear (Exercise 5.11).

Direct plug-in rules for selection of h are of the form

$$\hat{h}_{\text{DPI}} = C_1(K) \left[\frac{\hat{\sigma}_1^2(\lambda)}{\hat{\theta}_{22}(g)n} \right]^{1/5}.$$

As with all plug-in rules, we need to formulate rules for selection of the auxiliary bandwidths g and λ. Asymptotic approximations to the MSE-optimal choices of these bandwidths can be used to motivate their choice. Under sufficient smoothness assumptions the bandwidth that minimises $\text{MSE}\{\hat{\theta}_{22}(g)|X_1,\ldots,X_n\}$ is asymptotic to

$$g_{\text{AMSE}} = C_2(K) \left[\frac{\sigma^2}{|\theta_{24}|n} \right]^{1/7}$$

where $C_2(K) = \{12R(K_{2,3})/\mu_4(K_{2,3})\}^{1/7}$ if $\theta_{24} < 0$ and $C_2(K) = \{30R(K_{2,3})/\mu_4(K_{2,3})\}^{1/7}$ if $\theta_{24} > 0$. The bandwidth that minimises $\text{MSE}\{\hat{\sigma}_1^2(\lambda)|X_1,\ldots,X_n\}$ is asymptotic to

$$\lambda_{\text{AMSE}} = C_3(K) \left[\frac{\sigma^4}{\theta_{22}^2 n^2} \right]^{1/9}$$

where $C_3(K) = \{4R(K*K - 2K)/\mu_2(K)^4\}^{1/9}$ (Ruppert, Sheather and Wand, 1995).

As we saw in Section 3.5, one could replace θ_{24} and σ^2 by other kernel estimators, but this would lead to further bandwidth selection problems. A simple and effective strategy is to use a quick and simple estimate of m to estimate these parameters. One possibility is the "blocking method" (Härdle and Marron, 1995) where ordinary polynomial fits are obtained over each member of a partition of the range of the X data. The result is a relatively simple bandwidth selector for local linear regression that has been seen to work well in practice for a wide variety of functions. Also, it can be shown that, conditional on the X_i's, $\hat{h}_{\text{DPI}}/\hat{h}_{\text{MISE}} - 1 = O_P(n^{-2/7})$ (see Ruppert, Sheather and Wand, 1995). There are, of course, many other possible avenues for extension of the rules of Chapter 3, but we will not delve into these here.

5.9 Multivariate nonparametric regression

This chapter has so far been confined to the case of univariate design points. But as is clear from the huge literature on parametric regression, in very many practical applications we seek to identify how a response variable Y_i is related to d fixed design variables $\mathbf{x}_i = (x_{i1}, \ldots, x_{id})^T$ or random design variables $\mathbf{X}_i = (X_{i1}, \ldots, X_{id})^T$. This is the natural multivariate regression analogue of the multivariate density estimation problem treated in Chapter 4. Notationally, multivariate fixed and random design models are as in Section 5.1 except for the replacement of x_i and X_i by their multivariate extensions $\mathbf{x}_i$ and $\mathbf{X}_i$. Also, the design density f is now a d-variate function.

Fitting degree 1 polynomials locally will again be our focus in this section. A multivariate degree 1 polynomial is of the form $\beta_0 + \boldsymbol{\beta}_1^T \mathbf{x}$, a hyperplane depending on $d+1$ parameters β_0 and $\boldsymbol{\beta}_1 = (\beta_{11}, \ldots, \beta_{1d})^T$. Let K be a multivariate kernel function, satisfying condition (iii) of Section 4.3, and $\mathbf{H}$ be a bandwidth matrix as in Chapter 4. The multivariate regression analogue of the univariate local linear kernel estimator is

$$\hat{m}(\mathbf{x}; 1, \mathbf{H}) = \mathbf{e}_1^T (\mathbf{X}_\mathbf{x}^T \mathbf{W}_\mathbf{x} \mathbf{X}_\mathbf{x})^{-1} \mathbf{X}_\mathbf{x}^T \mathbf{W}_\mathbf{x} \mathbf{Y}$$

where

$$\mathbf{X}_\mathbf{x} = \begin{bmatrix} 1 & (\mathbf{X}_1 - \mathbf{x})^T \\ \vdots & \vdots \\ 1 & (\mathbf{X}_n - \mathbf{x})^T \end{bmatrix}$$

and

$$\mathbf{W}_\mathbf{x} = \text{diag}\{K_\mathbf{H}(\mathbf{X}_1 - \mathbf{x}), \ldots, K_\mathbf{H}(\mathbf{X}_n - \mathbf{x})\}.$$

Assuming that $\mathbf{H}$ satisfies condition (ii) of Section 4.3 and that $f(\mathbf{x})$, $v(\mathbf{x})$ and all entries of the Hessian $\mathcal{H}_m(\mathbf{x})$ are continuous, arguments analogous to those used in Section 4.3 and 5.3 lead to

$$\begin{aligned} &E\{\hat{m}(\mathbf{x}; 1, \mathbf{H}) - m(\mathbf{x}) | \mathbf{X}_1, \ldots, \mathbf{X}_n\} \\ &= \tfrac{1}{2}\mu_2(K)\text{tr}\{\mathbf{H}\mathcal{H}_m(\mathbf{x})\} + o_P\{\text{tr}(\mathbf{H})\} \end{aligned} \quad (5.21)$$

(Exercise 5.12) and

$$\begin{aligned} &\text{Var}\{\hat{m}(\mathbf{x}; 1, \mathbf{H}) | \mathbf{X}_1, \ldots, \mathbf{X}_n\} \\ &= n^{-1} |\mathbf{H}|^{-1/2} R(K) v(\mathbf{x})/f(\mathbf{x}) + o\{n^{-1}|\mathbf{H}|^{-1/2}\} \end{aligned}$$

(Ruppert and Wand, 1994). These conditional bias and variance expressions are the natural extensions of equations (5.9) and (5.10) in the same way that expressions (4.7) and (4.8) extend (2.10) and (2.11). They have the same simple interpretation as these other cases and can also be used as a basis for bandwidth selection by extending the ideas of the previous section.

Just as in density estimation, direct multivariate kernel estimators suffer from the curse of dimensionality. In the regression context a variety of alternative approaches has been proposed to overcome this problem. One of the simplest is the *additive modelling* methodology discussed in detail by Hastie and Tibshirani (1990). Here the general regression function is modelled to be of the form

$$\alpha + \sum_{j=1}^{d} m_j(x_j). \qquad (5.22)$$

That is, one simply models $m(\mathbf{x})$ as a sum of univariate functions, one for each coordinate direction. Kernel smoothing can be used to estimate each m_j. This is not the same as performing d ordinary univariate regressions since one must take the other dimensions into account when fitting in each direction. Iterative procedures known as *backfitting*, that involve residuals from current fits, are used to accomplish this. Estimation errors due to smoothing are essentially of a size connected with univariate estimation rather than direct multivariate estimation. Greater errors might arise, however, from the approximation of $m(\mathbf{x})$ by formula (5.22): general multivariate functions can be much more complicated than additive functions of individual coordinates. Many extensions of (5.22) are possible to combat this; see Ripley (1994, Section 3) for a succinct account.

5.10 Bibliographical notes

5.1 The Nadaraya-Watson kernel regression estimator was independently proposed by Nadaraya (1964) and Watson (1964). Priestley and Chao (1972) and Gasser and Müller (1979, 1984) introduced the alternative kernel regression estimators. Mack and Müller (1989) studied the properties of these alternative kernel estimators when applied to random designs. Eubank (1988), Müller (1988) and Härdle (1990b) contain extensive accounts of traditional kernel regression. See Altman (1992) for an introduction to kernel regression.

5.2 Local polynomial fitting has a long history in the smoothing of time series. Macauley (1931) is an early reference. Significant modern contributions to this methodology include Stone (1977), Cleveland (1979), Lejeune (1985), Müller (1987), Cleveland and Devlin (1988), Fan (1992a, 1993) and Fan and Gijbels (1992). Cleveland's proposal, based on nearest neighbour weights rather than kernel weights and with some robustification, also goes by the name *loess*. The work of Fan (1992a, 1993) is particularly noteworthy since it demonstrated the desirable MSE properties of $\hat{m}(x; 1, h)$, as well as establishing that local linear kernel estimators have certain minimax optimality properties. These important contributions have provoked a renewed interest in the methodology.

5.3 Results for the conditional mean squared error of $\hat{m}(x; 1, h)$ are due to Fan (1992a).

5.4–5.5 The formulae in these sections are due to Ruppert and Wand (1994). Fan and Gijbels (1992) first brought the attractive boundary properties of local least squares to prominence.

5.6 A detailed comparison of the properties of the Nadaraya-Watson and Gasser-Müller estimators (with discussion) is given by Chu and Marron (1991). Further work on how various kernel estimators deal with non-uniform designs is given by Jones, Davies and Park (1994). Lejeune (1985) and Müller (1987) studied certain equivalences between local polynomial kernel estimators and other kernel estimators. The effective kernel ideas used here are based on Hastie and Loader (1993). Earlier work along these lines may be found in Silverman (1984) and Hastie and Tibshirani (1990).

5.7 The results presented here are also from Ruppert and Wand (1994). Alternative estimators of derivatives were first developed by Gasser and Müller (1984). Jones (1994) compares the results of differentiating and of using coefficients of local polynomials for estimating derivatives.

5.8 The bandwidth selection algorithm presented here was formulated by Ruppert, Sheather and Wand (1995). There are several papers on bandwidth selection for the traditional kernel estimators. Härdle and Marron (1985) applied the ideas of least squares cross-validation to the Nadaraya-Watson estimator. An interesting study showing the poor asymptotic performance of cross-validatory selectors is given by Härdle, Hall and Marron (1988). Gasser, Kneip and Köhler (1991) and Härdle, Hall and Marron (1992) proposed bandwidth selectors with better asymptotic properties and practical performance. Härdle and Marron (1995) proposed quick and simple bandwidth selectors for the Nadaraya-Watson estimator.

5.9 The multivariate bias and variance calculations are also taken from Ruppert and Wand (1994). Hastie and Tibshirani (1990) contains a comprehensive account of additive modelling. Other approaches include projection pursuit regression (Friedman and Stuetzle, 1981), alternating conditional expectation (ACE) (Breiman and Friedman, 1985), average derivative estimation (Härdle and Stoker, 1989), sliced inverse regression (Li, 1991, Duan and Li, 1991) and multivariate adaptive regression splines (MARS) (Friedman, 1991); Ripley (1994, Section 3) mentions these and other similar methods, including neural networks.

5.11 Exercises

5.1 Let $(X_1, Y_1), \ldots, (X_n, Y_n)$ be a bivariate data set and suppose that the line $Y = \beta_0 + \beta_1 X$ is to be fitted using *weighted least squares*, that is to minimise

$$S = \sum_{i=1}^{n} (Y_i - \beta_0 - \beta_1 X_i)^2 w_i$$

for a set of weights $w_1, \ldots, w_n$.
(a) Show that
$$S = (\mathbf{Y} - \mathbf{X}\boldsymbol{\beta})^T \mathbf{W} (\mathbf{Y} - \mathbf{X}\boldsymbol{\beta})$$
where
$$\mathbf{Y} = \begin{bmatrix} Y_1 \\ \vdots \\ Y_n \end{bmatrix}, \quad \mathbf{X} = \begin{bmatrix} 1 & X_1 \\ \vdots & \vdots \\ 1 & X_n \end{bmatrix},$$
$\boldsymbol{\beta} = (\beta_0, \beta_1)^T$ and $\mathbf{W} = \text{diag}\{w_1, \ldots, w_n\}$.
(b) Let
$$\partial g(\boldsymbol{\beta})/\partial \boldsymbol{\beta} \equiv \begin{bmatrix} \partial g(\boldsymbol{\beta})/\partial \beta_0 \\ \partial g(\boldsymbol{\beta})/\partial \beta_1 \end{bmatrix}$$
where $g(\boldsymbol{\beta})$ is a scalar-valued function of $\boldsymbol{\beta}$. If $\mathbf{a}$ is a 2×1 vector and $\mathbf{A}$ is a 2×2 symmetric matrix then show that
$$\partial (\mathbf{a}^T \boldsymbol{\beta})/\partial \boldsymbol{\beta} = \mathbf{a} \quad \text{and} \quad \partial (\boldsymbol{\beta}^T \mathbf{A} \boldsymbol{\beta})/\partial \boldsymbol{\beta} = 2\mathbf{A}\boldsymbol{\beta}.$$

(c) Assuming that $\mathbf{X}^T \mathbf{W} \mathbf{X}$ is invertible show that S has a stationary point at
$$\hat{\boldsymbol{\beta}} = (\mathbf{X}^T \mathbf{W} \mathbf{X})^{-1} \mathbf{X}^T \mathbf{W} \mathbf{Y}.$$

Further analysis can be used to show that if $\mathbf{X}^T\mathbf{W}\mathbf{X}$ is positive definite then $\hat{\boldsymbol{\beta}}$ corresponds to the minimum of S and is therefore the solution to the weighted least squares linear regression problem.

5.2 Let $\hat{s}_r(x;h) = n^{-1}\sum_{i=1}^n (x_i - x)^r K_h(x_i - x)$ and $\hat{t}_r(x;h) = n^{-1}\sum_{i=1}^n (x_i - x)^r K_h(x_i - x)Y_i$.

(a) Show that

$$\hat{m}(x;p,h) = \mathbf{e}_1^T \begin{bmatrix} \hat{s}_0(x;h) & \cdots & \hat{s}_p(x;h) \\ \vdots & \ddots & \vdots \\ \hat{s}_p(x;h) & \cdots & \hat{s}_{2p}(x;h) \end{bmatrix}^{-1} \begin{bmatrix} \hat{t}_0(x;h) \\ \vdots \\ \hat{t}_p(x;h) \end{bmatrix}$$

(b) Using (a), derive expressions (5.3) and (5.4).

5.3 Let $p = 1$ and define

$$K^*(u;x) = \mathbf{e}_1^T (\mathbf{X}_x^T \mathbf{W}_x \mathbf{X}_x)^{-1} [1 \ \ (u-x)]^T K_h(u-x).$$

(a) Show that $\hat{m}(x;h) = \sum_{i=1}^n K^*(X_i;x) Y_i$.
(b) Show that

$$\sum_{i=1}^n K^*(X_i;x) = 1 \quad \text{and} \quad \sum_{i=1}^n K^*(X_i;x)(X_i - x) = 0.$$

5.4 Show that $K_{(p)}$ given by (5.11) is a $(p+1)$th-order kernel if p is odd.

5.5 Suppose that $(X_1, Y_1), \ldots, (X_n, Y_n)$ is a random sample from the bivariate density f_{XY} and let f_X denote the marginal density of the X_i's. Note that

$$m(x) = E(Y|X = x) = \int y f_{XY}(x,y)\, dy \Big/ f_X(x).$$

Show that if $f_X(x)$ is replaced by the kernel density estimator having kernel K and bandwidth h and if f_{XY} is replaced by a bivariate kernel estimate with a product kernel based on K and bandwidth h used in the X direction then the Nadaraya-Watson estimator $\hat{m}_{NW}(x;h)$ results (Watson, 1964).

5.6 Verify that the formulae (5.12) and (5.13) hold in the case $p = 0$: that is, for the Nadaraya-Watson estimator.

5.7

(a) Compute an expression for

$$\int_{-\alpha}^{1} K_{(1)}(u,\alpha)^2 \, du \Big/ \int_{-\alpha}^{1} K_{(0)}(u,\alpha)^2 \, du$$

when $K(x) = \tfrac{1}{2} 1_{\{|x|<1\}}$ is the uniform kernel and $0 \leq \alpha \leq 1$.

(b) Calculate this ratio for $\alpha = 0, 0.25, 0.5, 0.75, 1$ and plot the results. Interpret the results in terms of the relative variances of the local linear and Nadaraya-Watson estimators based on K for x near the boundary.

5.8 Verify that $K_{(p)}(\cdot, \alpha)$ as defined by (5.15) satisfies the conditions of a $(p+1)$th-order boundary kernel whenever p is odd.

5.9 Obtain and compare the conditional asymptotic biases and variances of $\hat{m}_1(x;h)$ and $\hat{m}_2(x;h)$ given by (5.17) and (5.18) respectively.

5.10 Suppose that $\text{Var}(Y_i|X_1,\ldots,X_n) = \sigma^2$ for all i. Show that as $g \to 0$,

$$E\{\hat{\theta}_{22}(g) - \theta_{22}|X_1,\ldots,X_n\} = \tfrac{1}{12} g^2 \mu_4(K_{(2,3)}) \theta_{24} \\ + n^{-1} g^{-5} R(K_{(2,3)}) \sigma^2 + o_P(g^2).$$

5.11 Suppose that $\text{Var}(Y_i|X_1,\ldots,X_n) = \sigma^2$ for all i. Show that $\hat{\sigma}^2(\lambda)$ defined by (5.20) is exactly conditionally unbiased for σ^2 if m is linear.

5.12 Using arguments similar to those given in Sections 4.3 and 5.3 derive result (5.21).

CHAPTER 6

Selected extra topics

6.1 Introduction

Density estimation and nonparametric regression are the most fundamental and familiar problems where kernel smoothing techniques provide an effective solution. However, the principles of kernel smoothing can be generalised and adapted to overcome several more complicated problems. These can broadly be divided into:
- situations where the data do not conform with the usual random sample assumptions, such as those where the data have dependencies, or are observed with error;
- estimation of other functions. Examples include hazard rates, spectral densities and intensity functions.

In this chapter we study a selection of topics that describe some of these extensions.

In Section 6.2 we deal with the first type of extension, within the density estimation setting, and show how various types of corruption of the data influence the performance of suitably modified kernel density estimates.

The remaining sections each deal with a different curve estimation problem in which kernel smoothing ideas apply. Section 6.3 treats the hazard rate estimation problem, Section 6.4 deals with spectral density estimation. The extension of kernel regression ideas to likelihood-based methods is described in Section 6.5. Estimation of the intensity function of a non-homogeneous Poisson process is discussed in Section 6.6. In each of these cases we refrain from delving deeply into the mathematical properties of each estimator, but retain our goal of developing intuition into how each estimator works.

6.2 Kernel density estimation in other settings

6.2.1 Dependent data

So far in this book all analyses of kernel estimators have relied on the assumption of independence. While this assumption has led to invaluable understanding of the properties of kernel smoothers with fairly simple mathematical arguments, it should be acknowledged that there are many practical situations where the independence assumption is tenuous. A comprehensive treatment of the analyses of kernel estimators under various dependence assumptions is beyond the scope of this book. Nevertheless, it is worthwhile to consider some simple cases to give a feeling for what is involved.

Consider the kernel density estimator

$$\hat{f}(x;h) = n^{-1}\sum_{i=1}^{n} K_h(x - X_i)$$

where $X_1, \ldots, X_n$ are not necessarily independent, but belong to a strictly stationary process $\{X_j : -\infty < j < \infty\}$. By strict stationarity we mean that the X_j are identically distributed with common density f such that $\text{Cov}(X_j, X_{j+k})$ depends only on k. The bias of $\hat{f}(x;h)$ is unaffected by dependence since

$$E\hat{f}(x;h) = EK_h(x - X_1) = (K_h * f)(x)$$

as for the independent data case. However

$$\text{Var}\hat{f}(x;h) = n^{-1}\text{Var}K_h(x - X_1)$$
$$+ 2n^{-1}\sum_{j=1}^{n-1}(1 - j/n)\text{Cov}\{K_h(x - X_1), K_h(x - X_{j+1})\} \quad (6.1)$$

because of stationarity (Exercise 6.1). The first term on the right hand side of (6.1) is equal to the variance of the kernel density estimator based on independent data. The second term reflects the extra variability in $\hat{f}(x;h)$ due to the dependence of the X_i's.

One can apply the usual asymptotic theory to $\text{Var}\{\hat{f}(\cdot;h)\}$ and combine the result with the bias approximation from (2.10) to obtain an approximation for $\text{MISE}\{\hat{f}(\cdot;h)\}$. However, the results heavily depend on the strength of the dependence among the X_j's. A simple type of dependence, which covers a wide range of situations, is *moving average* or *linear* dependence for which the X_j's satisfy

$$X_j = \mu + \sum_{k=-\infty}^{\infty} a_k Z_{j-k} \qquad (6.2)$$

where $\{Z_j : -\infty < j < \infty\}$ are independent and identically distributed random variables with zero mean and variance σ^2, and $\sum_k a_k^2 < \infty$. The covariance function of the process is given by

$$r(j) = \text{Cov}(X_0, X_j) = \sigma^2 \sum_k a_k a_{k-j}.$$

The strength of the dependence can be measured through the size of $\sum_j r(j)$. If $\sum_j r(j) < \infty$ then the process is usually said to be *short-range* dependent, while *long-range* dependence is equivalent to $\sum_j r(j)$ being divergent. For data satisfying (6.2) it can be shown that, under the usual conditions on the bandwidth and some additional regularity assumptions,

$$\int \text{Var}\{\hat{f}(x;h)\}\,dx \sim (nh)^{-1} R(K) + \text{Var}(\overline{X}) R(f')$$

where $\overline{X} = n^{-1} \sum_{i=1}^n X_i$ is the sample mean (Hall and Hart, 1990). For short-range dependence a standard result is

$$\text{Var}(\overline{X}) = O(n^{-1}) = o\{(nh)^{-1}\}$$

so

$$\text{MISE}\{\hat{f}(\cdot;h)\} = (nh)^{-1} R(K) + \tfrac{1}{4} h^4 \mu_2(K)^2 R(f'') \\ + o\{(nh)^{-1} + h^4\}, \qquad (6.3)$$

as for independent data. However, for long-range dependent data the asymptotic size of $\text{Var}(\overline{X})$ can be significantly larger and can dominate the usual leading variance term, leading to a worse rate of convergence of $\hat{f}(\cdot;h)$. For example, if the covariance function satisfies

$$r(j) = C j^{-\alpha}, \qquad 0 < \alpha < 1$$

for some constant $C > 0$ then the optimal rate of convergence of $\text{MISE}\{\hat{f}(\cdot;h)\}$ is of order $n^{-\min\{4/5, \alpha\}}$.

6.2 KERNEL DENSITY ESTIMATION IN OTHER SETTINGS

Results of type (6.3), where the dependence is weak enough not to affect the leading term of the MISE approximation, are quite common in the literature (see e.g. Györfi, Härdle, Sarda and Vieu, 1990). For example, kernel estimators based on data from a finite moving average process, or any m-dependent data set, have the same leading term performance as those based on independent samples. Needless to say, such results need to be viewed with caution since dependencies in the data are certain to have some effect on the performance of $\hat{f}(\cdot; h)$, since they usually represent loss of information. One way of measuring the finite sample cost of dependence is through exact MISE analysis. Let φ_f and φ_K denote the characteristic functions of f and K and let $\mathrm{Re}(z)$ denote the real part of the complex number z. Then one can show that

$$\mathrm{MISE}\{\hat{f}(\cdot; h)\} = \mathrm{MISE}_0\{\hat{f}(\cdot; h)\}$$
$$+ (\pi n)^{-1} \sum_{j=1}^{n} (1 - j/n) \int |\varphi_K(ht)|^2 \qquad (6.4)$$
$$\times \mathrm{Re}[E \exp\{it(X_{j+1} - X_1)\} - |\varphi_f(t)|^2] \, dt$$

where $\mathrm{MISE}_0\{\hat{f}(\cdot; h)\}$ is the MISE of $\hat{f}(\cdot; h)$ if it were based on a sample of independent data (given by (2.8)). This formula can be difficult to handle in general, but can be simplified in particular special cases.

EXAMPLE. Suppose that $X_1, \ldots, X_n$ are generated by a first-order autoregressive (AR(1)) process:

$$X_j = \rho X_{j-1} + (1 - \rho^2)^{1/2} Z_j, \quad |\rho| < 1$$

where the Z_j are independent $N(0, 1)$ random variables. Then $X_1, \ldots, X_n$ are a *dependent* sample from the $N(0, 1)$ distribution. If this process is inverted and written in moving average form then we have $a_j = \rho^j$ and $r(j) = \rho^j$, $j \geq 0$, so these data exhibit short-range dependence and the asymptotic MISE is not affected by the dependence. For the exact MISE we obtain from (6.4)

$$\mathrm{MISE}\{\hat{f}(\cdot; h)\} = \mathrm{MISE}_0\{\hat{f}(\cdot; h)\} + (\pi n)^{-1} \sum_{j=1}^{n} (1 - j/n)$$
$$\times \{(1 - \rho^j + h^2)^{-1/2} - (1 + h^2)^{-1/2}\} \qquad (6.5)$$

(see Exercise 6.2). Figure 6.1 shows $\mathrm{MISE}\{\hat{f}(\cdot; h)\}$ versus $\log_{10}(h)$ for independent data ($\rho = 0$) (solid curve), $\rho = 0.5$ (dashed curve) and $\rho = 0.9$ (dot-dashed curve) when $n = 50$ and the kernel is the standard normal. The curves for $\rho = 0$ and $\rho = 0.5$ look similar, but notice that the latter's minimum MISE is about 50% larger than that for independent data, which shows that there is a price to be paid for having this much dependence. In fact, it can be calculated that 70% more data, $n = 85$, are required to achieve the same minimum MISE as for independent data. The cost of dependence is much more dramatic for $\rho = 0.9$, as shown by its much higher MISE curve. Almost ten times as many data, $n = 486$, are required to achieve the same minimum MISE as for independent data. Results of this type are to be expected since, as ρ gets closer to 1, the amount of information in the data reduces to that of a single observation.

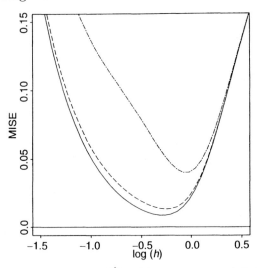

Figure 6.1. *Plots of* MISE$\{\hat{f}(\cdot\,; h)\}$ *versus* $\log_{10}(h)$ *for the AR(1) model with* $\rho = 0.5$ *(dashed curve) and* $\rho = 0.9$ *(dot-dashed curve). The solid curve is* MISE$\{\hat{f}(\cdot\,; h)\}$ *for independent data. The sample size is* $n = 50$ *and the kernel is the standard normal.*

■

6.2.2 Length biased data

Length biased data sets arise when the sampling mechanism is such that the larger the potential observation, the higher the probability it has of being included in the final sample: the probability of retention of a random draw X from a density f is proportional

6.2 KERNEL DENSITY ESTIMATION IN OTHER SETTINGS

to the value of X. This occurs in practice if one arrives at some ongoing process of independent and identically distributed "inter-event times" at some random point in time and observes the current inter-event length: we are more likely to arrive at some point in a longer lifetime than to hit a shorter one.

The problem of density estimation from length biased data can be formulated as follows: we observe a sample $X_1, \ldots, X_n$ of positive-valued random variables having density

$$g(x) = xf(x)/\mu, \quad x > 0 \qquad (6.6)$$

where $\mu = \int zf(z)dz < \infty$, and it is f that we wish to estimate. One could proceed by obtaining a kernel estimate $\hat{g}(x;h)$ of $g(x)$ and multiplying by $\hat{\mu}/x$, where $\hat{\mu}$ is an appropriate estimate of μ. But this method has some disadvantages. A notable one is that if $\hat{g}(x;h)$ does not go to zero as $x \to 0$ then the estimate tends to ∞ there. We prefer to take an alternative route. Let us replace the empirical distribution function $\hat{F}(x) = n^{-1} \sum_{i=1}^{n} 1_{\{X_i \leq x\}}$ by one that takes the length biased sampling mechanism into account. Intuitively, to redress the length biasing, make the steps of the empirical function inversely proportional to X_i; that is, take

$$\hat{F}_L(x) = n^{-1} \hat{\mu} \sum_{i=1}^{n} X_i^{-1} 1_{\{X_i \leq x\}},$$

where

$$\hat{\mu} = \left(n^{-1} \sum_{i=1}^{n} X_i^{-1} \right)^{-1}$$

is a root-n consistent estimator for μ (e.g. Cox, 1969). $\hat{F}_L$ is shown for a length biased data set of 46 widths of shrubs (source: Muttlak and McDonald, 1990) in Figure 6.2. Note the decreasing step sizes as x increases. It is easy to show that $\hat{F}_L$ is unbiased for F; indeed, $\hat{F}_L$ has the attractive properties associated with being the "nonparametric maximum likelihood estimator" of F in this context (Vardi, 1982).

To smooth, convolve $\hat{F}_L$ with K_h and differentiate (or, equivalently, convolve K_h with the empirical density function associated with $\hat{F}_L$, namely, $n^{-1}\hat{\mu} \sum_{i=1}^{n} X_i^{-1} \delta(x - X_i)$, where δ is the Dirac delta function). The result is

$$\hat{f}_L(x;h) = \int K_h(x-y)d\hat{F}_L(y)$$
$$= n^{-1}\hat{\mu}\sum_{i=1}^{n} X_i^{-1}K_h(x-X_i). \quad (6.7)$$

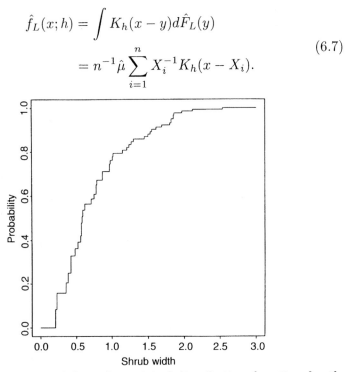

Figure 6.2. *Length biased empirical distribution function for the shrub width data.*

Note that the kernel density estimator for ordinary data can be written as

$$\hat{f}(x;h) = \int K_h(x-y)d\hat{F}(y). \quad (6.8)$$

Therefore (6.7) and (6.8) use the same idea of convolving a kernel weight with the density estimate induced by the natural estimate of the distribution function. The estimator $\hat{f}_L(x;h)$ has attractive mean squared error properties. If μ is known, these are very easy to obtain, and so, for pedagogical reasons, we treat this case here. The asymptotic approximations are the same when μ is replaced by an estimate. This is because $\hat{\mu}$ differs from μ by terms of order n^{-1}, so in asymptotic terms, estimation of μ is easy compared with the density estimation problem.

With μ known we immediately see that

$$E\{\hat{f}_L(x;h)\} = \mu(K_h * (x^{-1}g))(x) = (K_h * f)(x).$$

(Here, $x^{-1}g$ is shorthand for the function $g(x)/x$.) Therefore, $\hat{f}_L(x;h)$ has the same bias properties as the kernel density estimator based on ordinary data, including the dependence of the leading bias term on $f''(x)$.

By similar arguments, the leading variance term is

$$\operatorname{Var}\{\hat{f}_L(x;h)\} = n^{-1}\mu^2(K_h^2 * (x^{-2}g))(x)$$
$$= n^{-1}(K_h^2 * (\mu x^{-1}f))(x).$$

The integrated variance approximation is therefore

$$\int \operatorname{Var}\{\hat{f}_L(x;h)\}\, dx \sim (nh)^{-1}R(K)\mu\bar{\mu}$$

where $\bar{\mu} = \int z^{-1}f(z)dz$, which we assume to be finite. It follows from Jensen's inequality that $\mu\bar{\mu} \geq 1$, so the asymptotic integrated variance of $\hat{f}_L(\cdot;h)$ exceeds that of the kernel estimator based on ordinary data. This reflects the fact that we necessarily lose some information through the length biasing mechanism.

Figure 6.3 shows the effect of length biasing on density estimation. The dashed line is the kernel estimate for Muttlak and McDonald's (1990) shrub width data ignoring length bias: it therefore estimates g. The solid line is our estimate of f. Observe how much the latter differs from the former, and in what way. Here $h = 0.23$ in each case and K is the standard normal kernel.

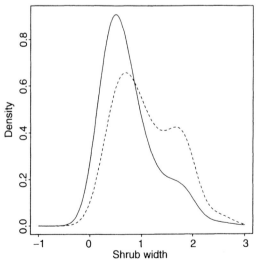

Figure 6.3. *Estimates of f and g for the shrub width data.*

At this point it is worth making some general comments about the appropriate choice of kernel estimator when the data, in some sense, pertain more directly to some curve other than the one of interest. In the case of length biased data this involves choosing between

$$\tilde{f}_L(x;h) = \hat{\mu}\hat{g}(x;h)/x$$

and

$$\hat{f}_L(x;h) = \int K_h(x-y)d\hat{F}_L(y).$$

Later in this chapter we will see that this dichotomy is present in other curve estimation settings, such as hazard rate estimation. The two approaches can be broadly described as

(i) smooth first and then adapt to the situation at hand, and
(ii) adapt to the situation at hand and then smooth.

While (i) is sometimes thought to be the more obvious approach it is approach (ii) that leads to the more appealing estimator. Appealing features include simpler interpretation, immediate adaptation to derivative estimation, simpler bias expressions and useful martingale representations (Jones, 1991b, Patil, Wells and Marron, 1995).

6.2.3 Right-censored data

Right-censored data typically arise in studies involving lifetimes, when there is a possibility that an item is removed from the study before the end of its "natural" lifetime. For example, if the lifetimes of a set of lightbulbs are being observed, then one that is accidentally broken before it burns out is said to have its natural lifetime *censored from the right* (see e.g. Cox and Oakes, 1984).

To put right-censorship into a mathematical framework we will let $X_1, \ldots, X_n$ denote the uncensored lifetimes and $Z_1, \ldots, Z_n$ denote the *censoring variables*, which we assume to be a random sample independent of the X_i's. Let F_X and F_Z denote the distribution functions of the X_i's and Z_i's, respectively. Rather than observe the X_i's we observe

$$Y_i = \min\{X_i, Z_i\} \quad \text{and} \quad I_i = 1_{\{X_i \leq Z_i\}}, \quad i = 1, \ldots, n,$$

that is, either the censored or completed lifetime together with knowledge of whether or not the lifetime we observe is a censored one. Our goal is to estimate f_X, the density of the X_i's.

A kernel density estimator of f_X can be motivated through the *Kaplan-Meier estimator* of F_X (Kaplan and Meier, 1958), which

6.2 KERNEL DENSITY ESTIMATION IN OTHER SETTINGS

is a generalization of the empirical distribution function for right-censored samples. The Kaplan-Meier estimate is given by

$$\hat{F}_X^{KM}(x) = \begin{cases} 0, & 0 \leq x \leq Y_{(1)} \\ 1 - \prod_{i=1}^{j-1} \left(\frac{n-i}{n-i+1}\right)^{I_{[i]}}, & Y_{(j-1)} < x \leq Y_{(j)} \\ & j = 2, \ldots, n, \\ 1, & x > Y_{(n)} \end{cases}$$

where $(Y_{(i)}, I_{[i]})$, $i = 1, \ldots, n$, denotes the (Y_i, I_i) ordered with respect to the Y_i's, and is the nonparametric maximum likelihood estimator of F_X under right-censorship. The kernel estimator of $f_X(x)$ induced by $\hat{F}_X^{KM}$ is then

$$\hat{f}_X(x; h) = \int K_h(x-y) d\hat{F}_X^{KM}(y)$$
$$= \sum_{i=1}^{n} s_i K_h(x - Y_{(i)}) \quad (6.9)$$

where s_i is the size of the jump of $\hat{F}_X^{KM}$ at $Y_{(i)}$. This estimator is based on the same principle as $\hat{f}_L(x; h)$ defined at (6.7), and shares the naturalness of that estimator (Patil, Wells and Marron, 1995).

It is easy to show that $s_i = 0$ if and only if $Y_{(i)}$ corresponds to a censored observation. Therefore, (6.9) uses only the uncensored X_i's. Figure 6.4 shows the estimate (6.9) for an artificial right-censored data set. The closed circles correspond to the uncensored observations, while the open circle denotes a censored data point. The kernel weights $s_i K_h(\cdot - X_i)$, $i = 1, \ldots, n$, correspond to the dashed curves while the resulting density estimate $\hat{f}_X(\cdot; h)$ corresponds to the solid curve. The Kaplan-Meier estimate $\hat{F}_X^{KM}$ is shown by the dotted lines. Note that the jumps in $\hat{F}_X^{KM}$ increase to the right of the censored data point since more weight is assigned to those observations that escape the censoring mechanism. This leads to an increase in kernel mass for the two right-most data points, reflecting the fact that the completed lifetime of the censored individual would have been to the right of its censored value.

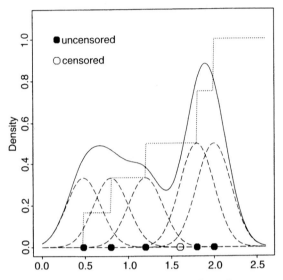

Figure 6.4. *Kernel density estimate based on a small right-censored data set (solid curve). Closed circles denote uncensored observations. The open circle is a censored observation. The dashed curves are the kernel weights; the dotted line is the Kaplan-Meier distribution function estimate.*

6.2.4 *Data measured with error*

Another problem of considerable practical interest is the nonparametric estimation of a curve from data that are measured with some non-negligible error. In this section we will focus on the estimation of a density from error contaminated data.

Suppose that $X_1, \ldots, X_n$ are a random sample having common density f_X, which we aim to estimate. However, the observed data are $Y_1, \ldots, Y_n$ where

$$Y_i = X_i + Z_i, \qquad i = 1, \ldots, n \qquad (6.10)$$

and, for each i, Z_i is a random variable that is independent of X_i and has known density f_Z, which we call the *error density*. If we apply the ordinary kernel estimate to the $Y_1, \ldots, Y_n$ then we will obtain a consistent estimate of

$$f_Y = f_X * f_Z$$

6.2 KERNEL DENSITY ESTIMATION IN OTHER SETTINGS

rather than f_X itself. Estimation of f_X requires that we take into account the fact that it is convolved with f_Z to give the density of the data. For this reason, estimation of f_X is often called the *deconvolution problem*.

A kernel-type solution to the deconvolution problem can be obtained by using Fourier transform (or characteristic function) properties. From (6.10) it is apparent that the characteristic function of each Y_i satisfies

$$\varphi_{f_Y}(t) = \varphi_{f_X}(t)\varphi_{f_Z}(t)$$

where φ_g is used to denote the characteristic function of a density g. According to the Fourier inversion theorem, the target density can be written as

$$f_X(x) = (2\pi)^{-1} \int e^{-itx} \varphi_{f_X}(t)\, dt$$

$$= (2\pi)^{-1} \int e^{-itx} \{\varphi_{f_Y}(t)/\varphi_{f_Z}(t)\}\, dt$$

provided $\varphi_{f_Z}(t) \neq 0$ for all t. An estimate of $f_X(x)$ is obtained by replacing f_Y by its kernel estimator $\hat{f}_Y(x; h) = n^{-1}\sum_{j=1}^{n} K_h(x - Y_j)$ to obtain

$$\hat{f}_X(x; h) = (2\pi)^{-1} \int e^{-itx} \{\varphi_{\hat{f}_Y(x;h)}(t)/\varphi_{f_Z}(t)\}\, dt, \qquad (6.11)$$

which we call the *deconvolving kernel density estimator* (Stefanski and Carroll, 1990). It can be shown (Exercise 6.5) that

$$\hat{f}_X(x; h) = n^{-1} \sum_{j=1}^{n} K_h^Z(x - Y_j; h) \qquad (6.12)$$

where $K_h^Z(u; h) = h^{-1} K^Z(u/h; h)$,

$$K^Z(u; h) = (2\pi)^{-1} \int e^{-itu} \{\varphi_K(t)/\varphi_{f_Z}(t/h)\}\, dt$$

and φ_K is the characteristic function of K. This shows that the deconvolving kernel density estimator has the same basic form as the ordinary kernel density estimator, but with K replaced by $K^Z(\cdot; h)$. This "effective" kernel differs from K in that its shape depends on the bandwidth, which is emphasised by having h as a second argument of K^Z.

EXAMPLE. Suppose that the error density is *Laplacian*, that is

$$f_Z(x) = (2\sigma)^{-1} e^{-|x|/\sigma}, \quad -\infty < x < \infty, \quad \sigma > 0$$

and that $K(x) = \phi(x)$, the standard normal kernel. Then, noting that the characteristic function of f_Z is

$$\varphi_{f_Z}(t) = (1 + \sigma^2 t^2)^{-1},$$

the effective kernel for deconvolution of Laplacian error can be shown to be

$$K^Z(x; h) = \phi(x)\{1 + (\sigma/h)^2(x^2 - 1)\}.$$

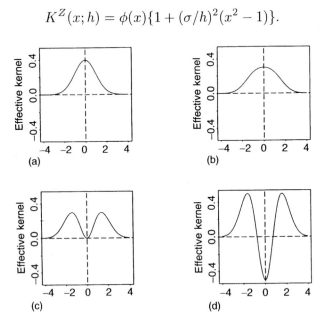

Figure 6.5. *Effective kernels $K^Z(\cdot; h)$ for deconvolution of Laplacian error. Values of σ/h are (a) 0, (b) $\frac{1}{2}$, (c) 1 and (d) $\frac{3}{2}$.*

Figure 6.5 shows the effective kernels for $\sigma/h = 0, \frac{1}{2}, 1, \frac{3}{2}$. Notice that, as the ratio σ/h increases, the effective kernel mass moves away from the origin. This allows for the fact that, for larger amounts of error, the probability of an observed value (X_i) lying close to its corresponding uncontaminated value (Y_i) diminishes.

6.2 KERNEL DENSITY ESTIMATION IN OTHER SETTINGS

In the fourth case, $\sigma/h = \frac{3}{2}$, the probability of X_i being away from Y_i is so great that the deconvolving kernel estimate assigns negative mass about Y_i. ∎

The MISE of the deconvolving density estimator can be shown to be

$$\text{MISE}\{\hat{f}_X(\cdot;h)\} = (nh)^{-1}R(K^Z(\cdot;h))$$
$$+ \int f_X(x)^2\,dx + (1-n^{-1})\int (K_h * f_X)^2(x)\,dx \quad (6.13)$$
$$- 2\int (K_h * f_X)(x) f_X(x)\,dx$$

(Exercise 6.6). Notice that this differs from the MISE for error-free observations only in the first term. In the error-free case this term is $(nh)^{-1}R(K)$. Therefore, the increase in $\text{MISE}\{\hat{f}_X(\cdot;h)\}$ due to having measurement error is

$$(nh)^{-1}\left\{R(K^Z(\cdot;h)) - R(K)\right\},$$

which depends only on the error density. Application of Parseval's identity leads to

$$R(K^Z(\cdot;h)) = (2\pi)^{-1}\int \varphi_K(t)^2|\varphi_{f_Z}(t/h)|^{-2}\,dt.$$

The effect of measurement error on $\text{MISE}\{\hat{f}_X(\cdot;h)\}$ is governed by the size of this integral which is, in turn, governed by the size of the reciprocal of the characteristic function of the error variable. For Laplacian error considered in the above example

$$\int \varphi_K(t)^2|\varphi_{f_Z}(t/h)|^{-2}\,dt = \int \varphi_K(t)^2\{1 + (\sigma/h)^2 t^2\}^2\,dt$$

which means that the leading integrated variance term is

$$\int \text{Var}\{\hat{f}_X(x;h)\}\,dx \sim (2\pi nh^5)^{-1}\sigma^4 \int t^4 \varphi_K(t)^2\,dt.$$

This can be traded off with the leading integrated squared bias term to get an optimal MISE rate of convergence of order $n^{-4/9}$ (see Exercise 6.7). Thus, having to deconvolve Laplacian error worsens the error-free MISE-optimal rate of order $n^{-4/5}$, reflecting less information in the sample.

Suppose instead that the error variable has a $N(0,\sigma^2)$ distribution, as is commonly assumed in practice. Then

$$\int \varphi_K(t)^2 |\varphi_{f_Z}(t/h)|^{-2}\, dt = \int \varphi_K(t)^2 e^{\sigma^2 t^2/h^2}\, dt.$$

The size of this integral can be extremely large and for most common kernels, including the normal if $h < \sigma$, this integral does not even converge. One way to overcome this problem is to take φ_K to have compact support, although this excludes many of the kernels used in practice (including all kernels listed in Table 2.1). Even with this restriction, it can be shown that the best obtainable rate of convergence of MISE$\{\hat{f}_X(\cdot;h)\}$ is of order $(\log n)^{-1}$, and that this lower bound applies to all deconvolving density estimators, not just (6.12) (Carroll and Hall, 1988). This very slow rate indicates that nonparametric curve estimation with normal errors can be extremely difficult. However, non-standard asymptotic analyses (Fan, 1992b) indicate that reasonably good density estimation may be possible for relatively small amounts of Gaussian error.

6.3 Hazard function estimation

A function of considerable importance in survival analysis and reliability is the *hazard function* (also called the *failure rate*) given by

$$\lambda(x) = f(x)/\{1 - F(x)\}, \qquad F(x) < 1$$

where f and F are, respectively, the density and distribution function of a positive-valued random variable X, typically representing the lifetime of a subject. The interpretation of the hazard function is through the fact that $\lambda(x)\,dx$ is the approximate probability of failure in the time interval $[x, x + dx]$, given that the subject has survived to time x (see e.g. McCune and McCune, 1987).

Based on the sample $X_1, \ldots, X_n$ of realizations of X, the kernel hazard function estimator of $\lambda(x)$ is

$$\hat{\lambda}(x;h) = n^{-1} \sum_{i=1}^{n} K_h(x - X_i)/\{1 - \widetilde{F}(X_i)\}$$

6.3 HAZARD FUNCTION ESTIMATION

where

$$\widetilde{F}(x) = n^{-1} \left\{ \sum_{i=1}^{n} 1_{\{X_i \leq x\}} - 1 \right\}$$

is a slight modification of the usual empirical distribution function which avoids division by zero at the maximum order statistic. There are a number of reasons why this version of the hazard function estimate, which again has the form

$$\int K_h(x-y) d\hat{F}(y)$$

for an appropriate "empirical cumulative hazard function" $\hat{F}$, is considered to be the most natural (Patil, Wells and Marron, 1995), see the discussion at the end of Section 6.2.2.

Figure 6.6 shows how $\lambda(\cdot; h)$ applies to an artificial data set of five observations. The dotted line shows the function $1 - \widetilde{F}$, while the dashed line shows the kernel weights at each X_i, which are inversely proportional to the height of $1 - \widetilde{F}(X_i)$. Therefore, the weight increases through the ordered X_i's, since the estimated probability of survival simultaneously decreases.

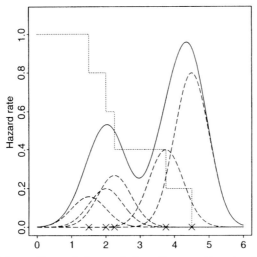

Figure 6.6. *Kernel hazard function estimate based on five observations (solid curve). Dashed curves show the kernel weight assigned to each observation. The dotted line is the adjusted empirical distribution function $\widetilde{F}$.*

6.4 Spectral density estimation

In this section we describe kernel estimation of the spectral density function. Before doing so, we note that it is rather ironic that spectral density estimation is considered so late in this book since it was developments in spectral smoothing that preceded and motivated the earliest works on probability density estimation. See Rosenblatt (1991) for this historical perspective.

Let $X(t)$ be a stationary process with respect to some time variable t. The *spectral density function* or *spectrum* is given by

$$g(\omega) = (2\pi)^{-1} \int_{-\pi}^{\pi} e^{-it\omega} r(t)\, dt$$

where $r(t) = \text{Cov}(X(0), X(t))$ is the covariance function of the process. Here ω represents *frequency*, and the spectrum is an important aspect of frequency domain analyses of time series data.

Based on a sample $X_1, \ldots, X_T$ from $X(t)$ at equally spaced intervals in time, the basic estimate of g is the *periodogram* $I_T(\omega)$ given by

$$I_T(\omega) = (2\pi T)^{-1} \left| \sum_{j=1}^{T} X_j e^{-ij\omega} \right|^2.$$

The periodogram derives from the Fourier transform of the sample autocovariances. However, the periodogram suffers from being a very wiggly estimator of the underlying continuous function. An example is given in Figure 6.7 (a) where the periodogram of the sunspot data (with sample mean subtracted), representing the annual means of daily relative sunspot numbers (source: Statistical Sciences, Inc., 1991), is plotted using a $\log_{10}$ vertical scale. In fact, if $X(t)$ is a zero mean stationary Gaussian process, it is a standard result of time series analysis that

$$I_T(\omega_1), \ldots, I_T(\omega_n)$$

behaves asymptotically (as $T \to \infty$) like a collection of independent scaled exponential random variables. It follows that $E\{I_T(\omega)\} \to g(\omega)$ and $\text{Var}\{I_T(\omega)\} \to g^2(\omega)$ (except at $\omega = 0$ and $\pm\pi$ where the asymptotic variance doubles). The main point is lack of consistency: the variance is fixed at $g^2(\omega)$ regardless of the size of T.

6.4 SPECTRAL DENSITY ESTIMATION

These deficiencies of the periodogram can be overcome by convolving $I_T(\omega)$ with a kernel weight K_h to obtain the *kernel spectral density estimator*

$$\hat{g}(\omega; h) = \int_{-\pi}^{\pi} K_h(\omega - u) I_T(u) du.$$

A kernel smoothed version of Figure 6.7 (a), using the standard normal kernel and $h = 0.10$, is displayed in Figure 6.7 (b).

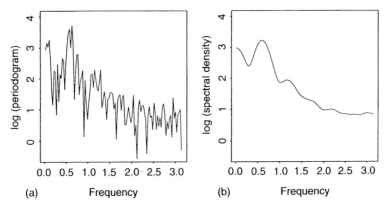

Figure 6.7. *(a) Periodogram and (b) kernel spectral density estimate with a normal kernel and a bandwidth of 0.10, each based on the sunspot data, and plotted using a $\log_{10}$ vertical scale.*

It can be shown that the bias of $\hat{g}(\omega; h)$ is dominated by

$$(K_h * g)(\omega) - g(\omega) \sim \tfrac{1}{2} h^2 \mu_2(K)^2 g''(\omega)$$

and its asymptotic variance can be shown to be

$$(nh)^{-1} R(K) 2\pi g^2(\omega)$$

(see Rosenblatt, 1991).

6.5 Likelihood-based regression models

An important recent extension of classical linear models is that to *generalised* linear models (see McCullagh and Nelder, 1988). Generalised linear models allow a variety of non-normal distributions for the regression response variable such as when the response is the number of successes from a binomial experiment or is a count variable.

The usual approach to estimation of regression coefficients in a generalised linear model is by maximum likelihood. This can be viewed as an extension of the standard regression problem with normal errors. To see this, consider the simple linear model

$$Y_i = \beta_0 + \beta_1 X_i + \varepsilon_i, \qquad i = 1, \ldots, n$$

where, conditional on the X_i's, the ε_i's are independent and identically distributed $N(0, \sigma^2)$ random variables. The conditional log-likelihood is

$$\ell = \sum_{i=1}^{n} \ln f_{Y|X}(Y_i|X_i)$$
$$= -\tfrac{1}{2}\sigma^{-2} \sum_{i=1}^{n} (Y_i - \beta_0 - \beta_1 X_i)^2 - \tfrac{n}{2} \ln(2\pi\sigma).$$

Therefore, choosing β_0 and β_1 to maximise ℓ is equivalent to choosing them by least squares. In Chapter 5 we dealt with the nonparametric extension of this approach using kernel-weighted least squares. In this section we demonstrate the adaptation of this extension for likelihood-based models.

For simplicity, we will first describe kernel-weighted likelihood-based methods in the familiar *binomial response* setting. In this case Y_i denotes the number of "successes" out of n_i trials at level X_i. Here the X_i's are realisations of a continuous variable, such as time or dosage. A typical example is one where we want to measure the toxicity of a certain drug. For $i = 1, \ldots, n$, dosages X_i of the drugs are administered to n_i experimental units, and the number of surviving units Y_i is recorded. The function of interest is

$$p(x) = P(\text{randomly selected unit survives at dosage } x).$$

Note that, conditional on X_i, Y_i is a binomial random variable with parameters n_i and $p(X_i)$. Therefore, we can write down the

6.5 LIKELIHOOD-BASED REGRESSION MODELS

conditional log-likelihood in terms of p:

$$\ell(p) = \sum_{i=1}^{n} [Y_i \ln\{p(X_i)\} + (n_i - Y_i) \ln\{1 - p(X_i)\}].$$

The usual parametric approach involves modelling p to satisfy

$$g\{p(x)\} = \beta_0 + \beta_1 x$$

where g is a one-to-one function from $(0, 1)$ to $(-\infty, \infty)$, usually called the *link* function. Use of a link function accounts for the fact that the usual "straight line" model is not appropriate for estimating probabilities. The parameters β_0 and β_1 can be chosen by maximum likelihood:

$$\begin{bmatrix} \hat{\beta}_0 \\ \hat{\beta}_1 \end{bmatrix} = \arg\max_{(\beta_0, \beta_1)^T} \sum_{i=1}^{n} [Y_i \ln\{g^{-1}(\beta_0 + \beta_1 X_i)\}$$
$$+ (n_i - Y_i) \ln\{1 - g^{-1}(\beta_0 + \beta_1 X_i)\}].$$

The estimate of $p(x)$ is then

$$\hat{p}(x) = g^{-1}(\hat{\beta}_0 + \hat{\beta}_1 x),$$

which is assured of being a probability by the properties of g. A popular choice of g is the logistic quantile function

$$g(u) = \text{logit}(u) = \ln\{u/(1-u)\}$$

in which case $\hat{p}$ is called the *logistic regression* estimate of p.

The extension of this approach to the nonparametric context by the introduction of kernel smoothing is straightforward. Instead of modelling $p(x) = g^{-1}(\beta_0 + \beta_1 x)$ *globally*, we do so *locally* by fitting β_0 and β_1 values for each x. This involves maximizing the log-likelihood suitably weighted by a kernel function. The estimate of $p(x)$ at x is

$$\hat{p}(x; h) = g^{-1}(\hat{\beta}_0) \tag{6.14}$$

where

$$\begin{bmatrix} \hat{\beta}_0 \\ \hat{\beta}_1 \end{bmatrix} = \arg\max_{(\beta_0, \beta_1)^T} \sum_{i=1}^{n} \Big(Y_i \ln\{g^{-1}(\beta_0 + \beta_1(X_i - x))\}$$
$$+ (n_i - Y_i)[\ln\{1 - g^{-1}(\beta_0 + \beta_1(X_i - x))\}]\Big) K_h(X_i - x).$$

Figure 6.8 shows the result of applying (6.14) to a set of birthweight data. The data represent proportions of surviving babies at 22 different birthweights. Interest centres on predicting survival based on birthweight (source: Karn and Penrose, 1951). Figure 6.8 (a) shows the estimate of logit$\{p(x)\}$, which is obtained by local linear fitting with respect to a kernel weight as depicted by the two kernels at the base of the picture. The kernel is the standard normal with bandwidth $h = 1.5$. This fitting process is analogous to that used for the local linear estimator of the ordinary regression function, discussed in Section 5.2. The estimate $\hat{p}(x;h)$, shown in Figure 6.8 (b) is obtained by applying the inverse logit transformation to Figure 6.8 (a).

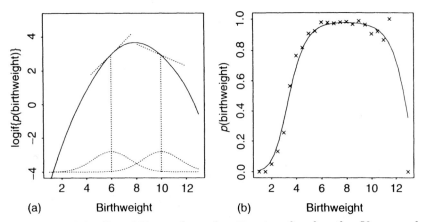

Figure 6.8. *Local linear kernel estimate of p for the Karn and Penrose birthweight data. The construction of the estimate of* logit$\{p(x)\}$ *is shown in (a). The estimate $\hat{p}(\cdot;h)$, shown in (b), is obtained from (a) by application of the inverse logit transformation. The crosses show the data.*

The extension to other likelihood-based settings is quite straightforward, as is the extension to fitting polynomials of arbitrary degree. An important class of likelihood-based models is those of exponential family form. Suppose that the conditional density of Y given $X = x$ can be written as

$$\ln f_{Y|X}(y|x) = y\theta(x) + b\{\theta(x)\} + c(y)$$

where b and c are known functions and

$$\theta(x) = (b')^{-1}(\mu(x)), \qquad \mu(x) = E(Y|X = x).$$

6.6 INTENSITY FUNCTION ESTIMATION

We usually call $\theta(x)$ the *canonical* or *natural* parameter and $g = (b')^{-1}$ is the canonical link function. The pth degree kernel-weighted polynomial estimate of $\mu(x)$ is

$$\hat{\mu}(x;h) = g^{-1}(\hat{\beta}_0)$$

where $\hat{\boldsymbol{\beta}} = (\hat{\beta}_0, \ldots, \hat{\beta}_p)^T$ and

$$\hat{\boldsymbol{\beta}} = \arg\max_{\boldsymbol{\beta}} \sum_{i=1}^{n} \Big(Y_i [g^{-1}\{\beta_0 + \ldots + \beta_p(X_i - x)^p\}]$$

$$+ b[g^{-1}\{\beta_0 + \ldots + \beta_p(X_i - x)^p\}] \Big) K_h(X_i - x).$$

As with ordinary local least squares regression, the interior and boundary bias are of the same order of magnitude, and asymptotically depend on x only through $\theta^{(p)}(x)$ when p is odd. For even p the boundary bias dominates the interior bias and both have a more complicated asymptotic expression (Fan, Heckman and Wand, 1995). This setup includes the special case $p = 0$, where it can be shown (Exercise 6.9) that the estimator for $\mu(x)$ reduces to

$$\hat{\mu}(x;h,0) = \sum_{i=1}^{n} Y_i K_h(X_i - x) \Big/ \sum_{i=1}^{n} K_h(X_i - x), \qquad (6.15)$$

the Nadaraya-Watson estimator. One could also replace the canonical link with other appropriate link functions and obtain analogous results.

Further extensions of the methodology presented here to quasi-likelihood-based models are also straightforward (Severini and Staniswalis, 1992, Fan, Heckman and Wand, 1995).

6.6 Intensity function estimation

Our final application of kernel smoothing ideas is to the estimation of the intensity function of a Poisson process.

Let $X_1, \ldots, X_N$ be ordered observations on the fixed interval $[0, T]$ of a non-homogeneous Poisson process having intensity function ρ. Then the natural kernel estimate of $\rho(x)$ is simply

$$\hat{\rho}(x;h) = \sum_{i=1}^{N} K_h(x - X_i)$$

(Ramlau-Hansen, 1983, Diggle, 1985). This is identical to the kernel density estimator except that there is no division by N, since $\rho(x)$ is based on an expected number rather than an expected proportion. It should also be noted that N is a random variable in this context.

Figure 6.9 shows a kernel intensity function estimate based on a set of data comprising the dates of major explosive volcanic eruptions in the Northern Hemisphere between 1851 and 1985 (source: Solow, 1991). In this case a normal kernel with a bandwidth of 10 years is used. Because of misleading boundary effects, the estimator is not shown within 15 years of each boundary. The dashed curve at the base of the figure shows a typical kernel function. The principles of kernel intensity functions are analogous to those of ordinary kernel density estimation, with kernel mass placed about each occurrence. The estimate shows periods of relatively low and high volcanic activity quite clearly.

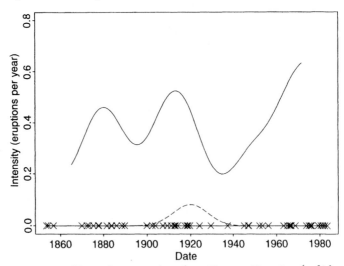

Figure 6.9. *Kernel intensity function estimate (solid curve) based on the volcanic eruption data. The estimate uses the normal kernel with a bandwidth of 10 years and is depicted by the dashed curve at the base of the picture.*

6.7 Bibliographical notes

6.2.1 There is a vast literature on the performance of kernel estimators under dependence. In lieu of a comprehensive survey of this work, we refer the reader to the monographs by Györfi, Härdle, Sarda and Vieu (1990) and Rosenblatt (1991). The asymptotic MISE results for infinite moving average processes given here are due to Hall and Hart (1990). Corresponding finite sample MISE results can be found in Hart (1984) and Wand (1992b). An investigation of least squares cross-validatory bandwidth selection in the presence of dependent data was given by Hart and Vieu (1990).

6.2.2 Cox (1969) and Vardi (1982) considered $\hat{F}_L$. The alternative kernel smoothing $\tilde{f}_L$ was proposed by Bhattacharyya, Franklin and Richardson (1988). The kernel smoother $\hat{f}_L$ was proposed and compared favourably with its rival by Jones (1991b).

6.2.4 The Kaplan-Meier estimator of the distribution function under right-censorship is based on Kaplan and Meier (1958). Early contributions to the theory of kernel density estimators based on the Kaplan-Meier estimator include Blum and Susarla (1980) and Földes, Rejtö and Winter (1981). A review of kernel density estimation from censored data is given by Padgett and McNichols (1984). Marron and Padgett (1987) address the bandwidth selection problem for right-censored data.

6.2.3 The deconvolving kernel density estimator presented here was developed and analysed by Stefanski and Carroll (1990). A related approach is due to Liu and Taylor (1989). Lower bounds on convergence rates for deconvolution were first obtained by Carroll and Hall (1988). Other noteworthy theoretical contributions are Devroye (1989), Fan (1991, 1992b) and Fan and Truong (1993).

6.3 The hazard rate function estimator described here was proposed by Watson and Leadbetter (1964). Contributions to hazard estimation include Rice and Rosenblatt (1976), Singpurwalla and Wong (1983), Tanner and Wong (1983, 1984), Uzunoğullari and Wang (1992) and Patil, Wells and Marron (1995). The last two references contains results on the naturalness of the hazard rate considered in this section.

6.4 See Rosenblatt (1991) for both historical development and an overview of modern work on spectral density estimation. Brillinger (1981, Chapter 5) also presents a detailed treatment of the problem. Bandwidth selection problems have been considered by Beltrão and Bloomfield (1987) and Park and Cho (1991).

6.5 The earliest approaches to smoothing in likelihood-based models were based on penalised likelihood ideas: see, for example, Silverman (1978), O'Sullivan, Yandell and Raynor (1986) and Green (1987). The notion of local likelihood was proposed by Tibshirani and Hastie (1987), and applied to the kernel estimation setting by Staniswalis (1989b) and Severini and Staniswalis (1992). Fan, Heckman and Wand (1995) extended the idea of local polynomial kernel regression to likelihood-based settings.

6.6 Kernel estimates of the intensity function were developed and analysed by Ramlau-Hansen (1983), Leadbetter and Wold (1983), Diggle (1985) and Diggle and Marron (1988). Brooks and Marron (1991) studied the bandwidth selection problem for this estimator. Further discussion on the volcanic eruption data may be found in Solow (1991).

6.8 Exercises

6.1 Derive result (6.1) for the variance of the kernel density estimator based on a stationary process.

6.2 Verify that (6.4) reduces to (6.5) for the first-order autoregressive dependence model.

6.3 An alternative to $\hat{f}_L(\cdot; h)$ for length biased data is

$$\tilde{f}_L(x; h) = n^{-1}(\hat{\mu}/x) \sum_{i=1}^{n} K_h(x - X_i).$$

Assuming that μ is known, obtain the mean and variance of $\tilde{f}_L(x; h)$.

6.4 Suppose that (6.6) were replaced by the more general relationship

$$g_W(x) = w(x)f(x)/\mu_W, \qquad x > 0,$$

where $\mu_W = \int w(z)f(z)\,dz < \infty$ and w is a known positive function. Obtain the natural extension of $\hat{f}_L(\cdot; h)$ to this situation and derive its asymptotic mean and variance, assuming that μ_W is known.

6.5 Show that the deconvolving kernel density estimator (6.11) can be rewritten as (6.12).

6.6 Verify formula (6.13) for the MISE of the deconvolving kernel density estimator.

6.8 EXERCISES

6.7 Show that for Laplacian error with scale parameter σ the AMISE-optimal bandwidth of the deconvolving kernel density estimator with second-order kernel K is

$$h_{\text{AMISE}} = \left[\frac{5\mu_4(\varphi_K^2)\sigma^4}{2\pi R(f'')\mu_2(K)^2 n}\right]^{1/9}$$

and the optimal mean integrated squared error is asymptotic to

$$\tfrac{9}{20}[625R(f'')^5\mu_2(K)^{10}\mu_4(\varphi_K^2)^4\{\sigma^4/(2\pi)\}]^{1/9}n^{-4/9}.$$

6.8 Suppose that, in the context of Section 6.2.4, the error density is the Cauchy:

$$f_Z(x) = (\pi\sigma)^{-1}\{1+(x/\sigma)^2\}^{-1}, \quad -\infty < x < \infty.$$

Derive an expression for the leading term of

$$\int \text{Var}\{\hat{f}_X(x;h)\}\,dx,$$

where $\hat{f}_X(\cdot;h)$ is based on a second-order kernel K.

6.9 Verify result (6.15). That is, show that the local constant ($p=0$) kernel-weighted estimator is the Nadaraya-Watson estimator for all exponential family models.

APPENDIX A

Notation

Notation concerning univariate functions

Let g be a real-valued univariate function.

$g_\lambda(x) = g(x/\lambda)/\lambda$ for $\lambda > 0$.

$\int g(x)\,dx = \int_{-\infty}^{\infty} g(x)\,dx$.

$g^{(r)}(x) = (d^r/dx^r)g(x)$.

$R(g) = \int g(x)^2\,dx$.

$\mu_\ell(g) = \int x^\ell g(x)\,dx$.

$(g * g_1)(x) = \int g(u)g_1(x-u)\,dx$, the convolution of g and g_1, where g_1 is another real-valued function.

$\varphi_f(t) = \int e^{itx} f(x)\,dx$, the characteristic function corresponding to the density f.

$\psi_r = \int f^{(r)}(x)f(x)\,dx$ for a density f and r an even integer.

$\phi(x) = (2\pi)^{-1/2} e^{-x^2/2}$, the density of the standard normal distribution.

$\phi_\sigma(x) = \phi(x/\sigma)/\sigma$, the density of the normal distribution having mean zero and variance σ^2.

$\Phi(x) = \int_{-\infty}^{x} \phi(t)\,dt$, the distribution function of the standard normal distribution.

$1_{\{x \in A\}} = 1$ for $x \in A$ and $1_{\{x \in A\}} = 0$ for $x \notin A$.

$\lfloor x \rfloor = $ greatest integer less than or equal to x.

Notation concerning multivariate functions

Let g be a real-valued d-variate function and $\mathbf{x} = (x_1, \ldots, x_d)$ be a d-vector.

NOTATION

$g_\Lambda(\mathbf{x}) = |\Lambda|^{-1/2} g(\Lambda^{-1/2}\mathbf{x})$ for Λ a symmetric and positive definite $d \times d$ matrix.

$\int g(\mathbf{x}) \, d\mathbf{x} = \int_{-\infty}^{\infty} \cdots \int_{-\infty}^{\infty} g(x_1, \ldots, x_d) \, dx_1 \cdots dx_d$.

$g^{(\mathbf{r})}(\mathbf{x}) = \partial^{|\mathbf{r}|}/(\partial x_1^{r_1} \cdots \partial x_d^{r_d}) g(\mathbf{x})$ where $\mathbf{r} = (r_1, \ldots, r_d)$ is a vector of non-negative integers.

$\mathcal{D}_g(\mathbf{x})$ is the $d \times 1$ vector with ith entry equal to $(\partial/\partial x_i)g(\mathbf{x})$, the derivative vector.

$\mathcal{H}_g(\mathbf{x})$ is the $d \times d$ matrix with (i,j) entry equal to $\partial^2/(\partial x_i \partial x_j)g(\mathbf{x})$, the Hessian matrix.

$\phi_\Sigma(\mathbf{x}) = (2\pi)^{-d/2} |\Sigma|^{-1/2} \exp(-\frac{1}{2}\mathbf{x}^T \Sigma^{-1} \mathbf{x})$, the density of the normal distribution with zero mean vector and covariance matrix Σ.

Notation concerning sequences

Let a_n and b_n be two real-valued deterministic sequences.

$a_n = o(b_n)$ as $n \to \infty$, if and only if $\lim_{n \to \infty} |a_n/b_n| = 0$.
$a_n = O(b_n)$ as $n \to \infty$, if and only if $\limsup_{n \to \infty} |a_n/b_n| < \infty$.
$a_n \sim b_n$, if and only if $\lim_{n \to \infty}(a_n/b_n) = 1$.

Let A_n and B_n be two real-valued random sequences.

$A_n = o_P(B_n)$ if for all $\varepsilon > 0$, $\lim_{n \to \infty} P(|A_n/B_n| > \varepsilon) = 0$.
$A_n = O_P(B_n)$ if for all $\varepsilon > 0$ there exist λ and M such that $P(|A_n/B_n| > \lambda) < \varepsilon$, for all $n > M$.
$A_n \to_D N(\mu, \sigma^2)$ if $\lim_{n \to \infty} P\{(A_n - \mu)/\sigma \leq x\} = \Phi(x)$ for all $x \in \mathbb{R}$; that is, A_n converges in distribution to a $N(\mu, \sigma^2)$ random variable.

Notation concerning vectors and matrices

Let $\mathbf{a} = (a_1, \ldots, a_d)$ be a d-vector and $\mathbf{A}$ a matrix.

$|\mathbf{a}| = a_1 + \ldots + a_d$.

diag($\mathbf{a}$) is the diagonal $d \times d$ matrix with the entries of $\mathbf{a}$ on the diagonal and zeros elsewhere.

tr($\mathbf{A}$) ($\mathbf{A}$ is a square matrix) is the trace of $\mathbf{A}$.

$|\mathbf{A}|$ ($\mathbf{A}$ is a square matrix) is the determinant of A.

vec $\mathbf{A}$ is the column vector obtained by stacking the columns of

A underneath each other in order from left to right, the vector of **A**.

vech **A** (**A** is a square matrix) is obtained from vec **A** by eliminating the above-diagonal entries of **A**, the vector-half of **A**.

$\mathbf{D}_d$ is the unique $d^2 \times \frac{1}{2}d(d+1)$ matrix of zeros and ones such that $\mathbf{D}_d$vech **A** = vec **A** for a symmetric $d \times d$ matrix **A**. $\mathbf{D}_d$ is called the duplication matrix of order d.

$\mathbf{e}_j$ denotes a column vector with jth entry equal to one and all other entries equal to zero.

I is the identity matrix.

Other notation and abbreviations

$\mathbb{R}$ is the real number line.

$\mathbb{R}^d$ is d-dimensional Euclidean space.

MSE: Mean Squared Error.

MISE: Mean Integrated Squared Error.

AMSE: Asymptotic Mean Squared Error.

AMISE: Asymptotic Mean Integrated Squared Error.

LSCV: Least Squares Cross-Validation.

BCV: Biased Cross-Validation.

DPI: Direct Plug-In.

STE: Solve-The-Equation.

SCV: Smoothed Cross-Validation.

APPENDIX B
Tables

Table B.1. *Probability densities and characteristic functions*

Name	Density function $f(x)$	Characteristic function $\varphi_f(t)$
Normal	$(2\pi)^{-1/2}e^{-x^2/2}$	$e^{-t^2/2}$
Cauchy	$\{\pi(1+x^2)\}^{-1}$	$e^{-|t|}$
Gamma(p)	$\Gamma(p)^{-1}x^{p-1}e^{-x}1_{\{x>0\}}$	$(1-it)^{-p}$
Laplace	$\frac{1}{2}e^{-|x|}$	$(1+t^2)^{-1}$
Uniform	$\frac{1}{2}1_{\{|x|<1\}}$	$(1/t)(\sin t)$
Epanechnikov	$\frac{3}{4}(1-x^2)1_{\{|x|<1\}}$	$(3/t^3)(\sin t - t\cos t)$
Biweight	$\frac{15}{16}(1-x^2)^2 1_{\{|x|<1\}}$	$(15/t^5)\{(3-t^2)\sin t - 3t\cos t\}$
Triweight	$\frac{35}{32}(1-x^2)^3 1_{\{|x|<1\}}$	$(105/t^7)\{(15-6t^2)\sin t - t(15-t^2)\cos t\}$
Triangular	$(1-|x|)1_{\{|x|<1\}}$	$(2/t^2)(1-\cos t)$
Logistic	$e^x/(e^x+1)^2$	$\pi t \operatorname{cosech}(\pi t)$
Extreme value	$e^x e^{-e^x}$	$\Gamma(1+it)$

Table B.2 *values of* $\int x^2 f(x)\,dx$, $\int f(x)^2\,dx$ *and* $\int f''(x)^2\,dx$

Density	$\int x^2 f(x)dx$	$\int f(x)^2 dx$	$\int f''(x)^2 dx$
Normal	1	$(2\pi^{1/2})^{-1}$	$3(8\pi^{1/2})^{-1}$
Cauchy	∞	$(2\pi)^{-1}$	$3(4\pi)^{-1}$
Gamma (1)	2	$\frac{1}{2}$	∞
Gamma (2)	6	$\frac{1}{4}$	∞
Gamma (3)	12	$\frac{3}{16}$	$\frac{3}{16}$
Gamma (4)	20	$\frac{5}{32}$	$\frac{1}{32}$
Laplace	2	$\frac{1}{4}$	∞
Uniform	$\frac{1}{3}$	$\frac{1}{2}$	∞
Epanechnikov	$\frac{1}{5}$	$\frac{3}{5}$	∞
Biweight	$\frac{1}{7}$	$\frac{5}{7}$	$\frac{45}{2}$
Triweight	$\frac{1}{9}$	$\frac{350}{429}$	35
Triangular	$\frac{1}{6}$	$\frac{2}{3}$	∞
Logistic	$\frac{1}{3}\pi^2$	$\frac{1}{6}$	$\frac{1}{42}$
Extreme Value	$\frac{1}{6}\pi^2 + \Gamma'(1)^2$	$\frac{1}{4}$	$\frac{1}{4}$

APPENDIX C
Facts about normal densities

C.1 Univariate normal densities

The standard normal probability density function is

$$\phi(x) = (2\pi)^{-1/2} e^{-x^2/2}.$$

A subscript means a rescaling of the type

$$\phi_\sigma(x) = \phi(x/\sigma)/\sigma.$$

Hence the $N(\mu, \sigma^2)$ density in the variable x is $\phi_\sigma(x - \mu)$.

The convention concerning derivatives and rescalings is that rescalings are done first, so

$$\phi_\sigma^{(r)}(x) = (d^r/dx^r)\phi_\sigma(x) = \phi^{(r)}(x/\sigma)/\sigma^{r+1}.$$

The rth Hermite polynomial is defined by

$$H_r(x) = (-1)^r \phi^{(r)}(x)/\phi(x).$$

For $r = 0, 1, \ldots$, the "Odd Factorial" is defined by

$$\text{OF}(2r) = (2r-1)(2r-3)\cdots 1 = \frac{(2r)!}{2^r r!}.$$

The characteristic function of a probability density function f is denoted by

$$\varphi_f(t) = \int e^{itx} f(x)\, dx.$$

177

Given functions f and g, the convolution of f and g (when it exists) is

$$(f * g)(x) = \int f(x-u)g(u)\,du.$$

The greatest integer less than or equal to x is denoted by $\lfloor x \rfloor$.

Fact C.1.1

$$H_0(x) = 1$$
$$H_1(x) = x$$
$$H_2(x) = x^2 - 1$$
$$H_3(x) = x^3 - 3x$$
$$H_4(x) = x^4 - 6x^2 + 3$$
$$H_5(x) = x^5 - 10x^3 + 15x$$
$$H_6(x) = x^6 - 15x^4 + 45x^2 - 15$$
$$H_7(x) = x^7 - 21x^5 + 105x^3 - 105x$$
$$H_8(x) = x^8 - 28x^6 + 210x^4 - 420x^2 + 105.$$

Fact C.1.2
$$\phi^{(r)}(x) = (-1)^r H_r(x)\phi(x).$$

Fact C.1.3
$$H_r(x) = \sum_{j=0}^{\lfloor r/2 \rfloor} (-1)^j \mathrm{OF}(2j) \binom{r}{2j} x^{r-2j}.$$

Fact C.1.4
$$H_r(x) = xH_{r-1}(x) - (r-1)H_{r-2}(x).$$

Fact C.1.5
$$(d/dx)H_r(x) = rH_{r-1}(x).$$

C.1 UNIVARIATE NORMAL DENSITIES

Fact C.1.6

$$\phi_\sigma^{(r)}(0) = \begin{cases} (-1)^{r/2}(2\pi)^{-1/2}\mathrm{OF}(r)\sigma^{-r-1} & r \text{ even} \\ 0 & r \text{ odd}. \end{cases}$$

Fact C.1.7 For $\sigma > 0$, $r = 0, 1, 2, \ldots$ and $X \sim N(\mu, \sigma^2)$,

$$E(X^r) = \sum_{j=0}^{\lfloor r/2 \rfloor} \mathrm{OF}(2j) \binom{r}{2j} \mu^{r-2j} \sigma^{2j}.$$

Fact C.1.8 For $X \sim N(0, \sigma^2)$, if r is even,

$$E(X^r) = \int x^r \phi_\sigma(x)\, dx = \sigma^r \mathrm{OF}(r),$$

and if r is odd

$$E(X^r) = \int x^r \phi_\sigma(x)\, dx = 0.$$

Fact C.1.9 For $\sigma, \sigma' > 0$,

$$\phi_\sigma(x - \mu)\phi_{\sigma'}(x - \mu')$$
$$= \phi_{(\sigma^2+\sigma'^2)^{1/2}}(\mu - \mu')\phi_{\sigma\sigma'/(\sigma^2+\sigma'^2)^{1/2}}(x - \mu^*)$$

where

$$\mu^* = \frac{\sigma'^2 \mu + \sigma^2 \mu'}{\sigma^2 + \sigma'^2}.$$

Fact C.1.10

$$\phi(x)^r = (2\pi)^{(1-r)/2} \phi_{r^{-1/2}}(x)/r^{1/2}.$$

Fact C.1.11 For $r_1, r_2, \ldots, r_m = 0, 1, 2, \ldots$

$$\phi_{\sigma_1}^{(r_1)}(\cdot - \mu_1) * \cdots * \phi_{\sigma_m}^{(r_m)}(\cdot - \mu_m)(x)$$
$$= \phi_{(\sigma_1^2+\ldots+\sigma_m^2)^{1/2}}^{(r_1+\ldots+r_m)}(x - \mu_1 - \ldots - \mu_m).$$

Fact C.1.12

$$\int \phi_\sigma^{(r)}(x-\mu)\phi_{\sigma'}^{(r')}(x-\mu')\,dx = (-1)^r \phi_{(\sigma^2+\sigma'^2)^{1/2}}^{(r+r')}(\mu-\mu').$$

Fact C.1.13

$$\varphi_{\phi_\sigma(\cdot-\mu)}(t) = 2\pi\phi_\sigma(\mu)\phi_{1/\sigma}(t-i\mu/\sigma^2) = e^{it\mu-\sigma^2 t^2/2}.$$

Fact C.1.14

$$\varphi_{\phi_\sigma}(t) = (2\pi)^{1/2}\phi_{1/\sigma}(t)/\sigma = e^{-\sigma^2 t^2/2}.$$

C.2 Multivariate normal densities

The standard d-variate normal probability density is

$$\phi(\mathbf{x}) = (2\pi)^{-d/2}\exp(-\tfrac{1}{2}\mathbf{x}^T\mathbf{x})$$

where $\mathbf{x} = (x_1,\ldots,x_d)^T$ is a vector in $\mathbb{R}^d$. Let $\boldsymbol{\Sigma}$ be a symmetric, positive definite $d\times d$ matrix. Then

$$\phi_{\boldsymbol{\Sigma}}(\mathbf{x}) = |\boldsymbol{\Sigma}|^{-1/2}\phi(\boldsymbol{\Sigma}^{-1/2}\mathbf{x})$$

so that $\phi_{\boldsymbol{\Sigma}}(\mathbf{x}-\boldsymbol{\mu})$ is the $N(\boldsymbol{\mu},\boldsymbol{\Sigma})$ density in the vector $\mathbf{x}$.

For a vector $\mathbf{r} = (r_1,\ldots,r_d)$ of non-negative integers define $|\mathbf{r}| = \sum_{i=1}^d r_i$. We will write multivariate derivatives of $\phi_{\boldsymbol{\Sigma}}(\mathbf{x})$ as

$$\phi_{\boldsymbol{\Sigma}}^{(\mathbf{r})}(\mathbf{x}) = \frac{\partial^{|\mathbf{r}|}}{\partial x_1^{r_1}\cdots \partial x_d^{r_d}}\phi_{\boldsymbol{\Sigma}}(\mathbf{x}).$$

The characteristic function of a d-variate density f is

$$\varphi_f(\mathbf{t}) = \int e^{i\mathbf{t}^T\mathbf{x}} f(\mathbf{x})\,d\mathbf{x}$$

and the convolution of two d-variate functions f and g is

$$(f*g)(\mathbf{x}) = \int f(\mathbf{x}-\mathbf{u})g(\mathbf{u})\,d\mathbf{u}.$$

Fact C.2.1 For any two multivariate normal distributions $N(\boldsymbol{\mu}, \boldsymbol{\Sigma})$ and $N(\boldsymbol{\mu}', \boldsymbol{\Sigma}')$

$$\phi_{\boldsymbol{\Sigma}}(\mathbf{x} - \boldsymbol{\mu})\phi_{\boldsymbol{\Sigma}'}(\mathbf{x} - \boldsymbol{\mu}') = \phi_{\boldsymbol{\Sigma}+\boldsymbol{\Sigma}'}(\boldsymbol{\mu} - \boldsymbol{\mu}')\phi_{\boldsymbol{\Sigma}(\boldsymbol{\Sigma}+\boldsymbol{\Sigma}')^{-1}\boldsymbol{\Sigma}'}(\mathbf{x} - \boldsymbol{\mu}^*)$$

where

$$\boldsymbol{\mu}^* = \boldsymbol{\Sigma}'(\boldsymbol{\Sigma} + \boldsymbol{\Sigma}')^{-1}\boldsymbol{\mu} + \boldsymbol{\Sigma}(\boldsymbol{\Sigma} + \boldsymbol{\Sigma}')^{-1}\boldsymbol{\mu}'.$$

Fact C.2.2 For vectors $\mathbf{r} = (r_1, \ldots, r_d)$ and $\mathbf{r}' = (r_1', \ldots, r_d')$ of non-negative integers

$$\phi_{\boldsymbol{\Sigma}}^{(\mathbf{r})}(\cdot - \boldsymbol{\mu}) * \phi_{\boldsymbol{\Sigma}'}^{(\mathbf{r}')}(\cdot - \boldsymbol{\mu}')(\mathbf{x}) = \phi_{\boldsymbol{\Sigma}+\boldsymbol{\Sigma}'}^{(\mathbf{r}+\mathbf{r}')}(\mathbf{x} - \boldsymbol{\mu} - \boldsymbol{\mu}').$$

Fact C.2.3

$$\int \phi_{\boldsymbol{\Sigma}}^{(\mathbf{r})}(\mathbf{x} - \boldsymbol{\mu})\phi_{\boldsymbol{\Sigma}'}^{(\mathbf{r}')}(\mathbf{x} - \boldsymbol{\mu}')\,d\mathbf{x} = (-1)^{|\mathbf{r}|}\phi_{\boldsymbol{\Sigma}+\boldsymbol{\Sigma}'}^{(\mathbf{r}+\mathbf{r}')}(\boldsymbol{\mu} - \boldsymbol{\mu}').$$

Fact C.2.4

$$\varphi_{\phi_{\boldsymbol{\Sigma}}(\cdot-\boldsymbol{\mu})}(\mathbf{t}) = (2\pi)^{d/2}e^{i\mathbf{t}^T\boldsymbol{\mu}}\phi_{\boldsymbol{\Sigma}^{-1}}(\mathbf{t})|\boldsymbol{\Sigma}|^{-1/2}.$$

Fact C.2.5

$$\varphi_{\phi_{\boldsymbol{\Sigma}}}(\mathbf{t}) = (2\pi)^{d/2}\phi_{\boldsymbol{\Sigma}^{-1}}(\mathbf{t})|\boldsymbol{\Sigma}|^{-1/2}.$$

C.3 Bibliographical notes

C.1 Proofs of the facts from this section may be found in Aldershof, Marron, Park and Wand (1992).

C.2 See Wand and Jones (1993) for the proofs of the main results in this section.

APPENDIX D

Computation of kernel estimators

D.1 Introduction

An important question that arises in the practical application of kernel estimators is that of their computation. Of course, since most kernel estimators have an explicit formulation one can always program kernel smoothing routines directly. However, direct computation can also be very computationally expensive. Consider, for example, the problem of obtaining a kernel density estimate over a mesh of M *grid points* $g_1, \ldots, g_M$ so that one can plot its graph. The direct approach is to use

$$\hat{f}_j \equiv \hat{f}(g_j; h) = n^{-1} \sum_{i=1}^{n} K_h(g_j - X_i), \quad j = 1, \ldots, M. \quad \text{(D.1)}$$

Notice that the number of kernel evaluations required to compute all of the $\hat{f}_j$ is nM. Other kernel estimators can pose even worse computational problems. For example, the kernel estimator of $\psi_m = \int f^{(m)}(x) f(x) \, dx$ given by

$$\hat{\psi}_m(h) = n^{-2} \sum_{i=1}^{n} \sum_{j=1}^{n} K_h^{(m)}(X_i - X_j)$$

requires $O(n^2)$ kernel evaluations which can make its computation very slow for even moderately large values of n.

In this appendix we will present one way of dramatically increasing the computational speed at the expense of replacing kernel estimators by approximations. The approximation is usually very good for moderate values of M, say between about 100 and 500 (Hall and Wand, 1994), and can be made arbitrarily good by increasing the value of M. In the density estimation case the main idea is to replace the data by *grid counts* $c_1, \ldots, c_M$, where c_j is a weight chosen to represent the amount of data near g_j. By computing a binned version of $\hat{f}(\cdot; h)$ based on the grid counts the number of kernel evaluations is only $O(M)$ which represents an immense saving, especially for large samples.

D.2 The binned kernel density estimator

The *binned kernel density estimator* (e.g. Silverman, 1982, Scott and Sheather, 1985, Härdle and Scott, 1992) provides a way of approximating the $\hat{f}_j$ as defined by (D.1). The approximation we use has a discrete convolution structure which can be computed quickly using the fast Fourier transform.

Let K be a symmetric kernel with support confined to $[-\tau, \tau]$ for $\tau > 0$. If K has infinite support then one can choose τ so that $[-\tau, \tau]$ is the "effective support" of K, that is the region outside which K is negligible. For the standard normal kernel $\tau = 4$ is a reasonable choice. Let $[a, b]$ be an interval containing all of the data and let $a = g_1 < g_1 < \cdots < g_M = b$ be an equally spaced grid on $[a, b]$. The binned kernel density estimator relies on assigning certain weights to the grid points, based on neighbouring observations, to obtain the grid counts $c_1, \ldots, c_M$. The exact specification of the c_ℓ will be given below. The binned kernel density estimator at a point x is given by

$$\tilde{f}(x; h, M) = n^{-1} \sum_{\ell=1}^{M} K_h(x - g_\ell) c_\ell.$$

It follows that our approximation to $\hat{f}_j$ is

$$\tilde{f}_j = n^{-1} \sum_{\ell=1}^{M} K_h(g_j - g_\ell) c_\ell, \quad j = 1, \ldots, M.$$

Noting that $c_\ell = 0$ for ℓ outside of the set $\{1, \ldots, M\}$ it is easily seen that

$$\tilde{f}_j = \sum_{\ell=1-M}^{M-1} c_{j-\ell} \kappa_\ell, \quad j = 1, \ldots, M \quad \text{(D.2)}$$

where
$$\kappa_\ell = n^{-1} K_h\left(\frac{(b-a)\ell}{M-1}\right), \quad \ell \text{ an integer.}$$

Formula (D.2) makes the gains that can be made by computing the $\tilde{f}_j$ rather than $\hat{f}_j$ very apparent. Firstly, by the symmetry of K it is clear that no more than M kernel evaluations are required to obtain the κ_ℓ. This is a consequence of the fact that there are only M distinct grid point differences as illustrated in Figure D.1.

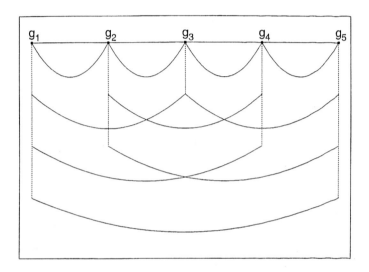

Figure D.1. *Illustration of distinct grid point differences for the case $M = 5$.*

The second noteworthy point is that the vector of $\tilde{f}_j$'s is the discrete convolution of the c_ℓ's and the κ_ℓ's so can be computed quickly using the *fast Fourier transform* which is an $O(M \log M)$ algorithm for computing discrete Fourier transforms. The details are given below.

The computation of the $\tilde{f}_j$ can be divided into three distinct steps:
- obtain the grid counts, c_ℓ, $\ell = 1, \ldots, M$,
- compute the kernel weights, κ_ℓ, for all integers ℓ such that $\kappa_\ell \neq 0$,
- weight the grid points by convolving the c_ℓ with the κ_ℓ.

Note that only the second and third stages need to be repeated if a kernel estimate with a different bandwidth is required for the same data set. The following three paragraphs describe each of these

D.2 THE BINNED KERNEL DENSITY ESTIMATOR

stages in turn.

Obtaining the grid counts

There are two commonly used strategies for obtaining the grid counts $c_1,\ldots,c_M$, a procedure usually referred to as *binning* or *discretising* the data. Suppose that an observation occurs at x and that the nearest grid points below and above x are at y and z respectively. Then the *simple binning* strategy assigns a unit weight to either y or z, whichever is closer. On the other hand, *linear binning* assigns the grid point at y a weight of $(z-x)/(z-y)$ and the grid point at z a weight of $(x-y)/(z-y)$. In terms of approximation error of $\hat{f}(\cdot;h)$ by $\tilde{f}(\cdot;h,M)$, linear binning is significantly more accurate than simple binning (Jones and Lotwick, 1983, Hall and Wand, 1994).

Computing the kernel weights

Since K is symmetric and compactly supported on $[-\tau,\tau]$ we only need to compute κ_ℓ for $\ell = 0, 1, \ldots, \lfloor \tau h(M-1)/(b-a)\rfloor$. Furthermore, (D.2) does not depend on those κ_ℓ for which $\ell \geq M$ so it suffices to compute κ_ℓ for $\ell = 0,\ldots,L$ where

$$L = \min\{\lfloor \tau h(M-1)/(b-a)\rfloor, M-1\}.$$

Performing the convolution

The convolution of the c_ℓ and κ_ℓ can be computed directly in $O(M^2)$ operations using

$$\tilde{f}_j = \sum_{\ell=-L}^{L} c_{j-\ell}\kappa_\ell, \quad j=1,\ldots,M$$

which typically represents a substantial saving over (D.2). One could also use Fourier transform methods to compute the required convolution. The *discrete Fourier transform* of a complex vector $\mathbf{z} = (z_0,\ldots,z_{N-1})$ is the vector $\mathbf{Z} = (Z_0,\ldots,Z_{N-1})$ where

$$Z_j = \sum_{\ell=0}^{N-1} z_\ell e^{2\pi i \ell j/N}, \quad j=0,\ldots,N-1.$$

The vector $\mathbf{z}$ can be recovered from its Fourier transform $\mathbf{Z}$ by applying the *inverse discrete Fourier transform* formula

$$z_\ell = N^{-1} \sum_{j=0}^{N-1} Z_j e^{-2\pi i \ell j/N}, \quad \ell=0,\ldots,N-1.$$

Discrete Fourier transforms and their inverses can be computed in $O(N \log N)$ operations using the fast Fourier transform (FFT) algorithm. The algorithm is fastest when N is highly composite, such as a power of 2 (see e.g. Press, Flannery, Teukolsky and Vetterling, 1988, Chapter 12). The discrete convolution of two vectors can be computed quickly using the FFT by appealing to the *discrete convolution theorem* (Press et al., 1988, p.408): multiply the Fourier transforms of the two vectors element-by-element and then invert the result to obtain the convolution vector. However, this theorem requires certain periodicity assumptions, so when these assumptions are violated appropriate *zero-padding* is required to avoid *wrap-around* effects (see Press et al., 1988, pp.410–411). We will give a description of what is required for the FFT computation of the convolution given at (D.2). Let P be a power of 2 such that $P \geq M + L$ and let $\mathbf{0}_p$ denote the $1 \times p$ vector of zeros. Define

$$\mathbf{c} = (c_1, \ldots, c_M, \mathbf{0}_{P-M}) \quad \text{and}$$

$$\boldsymbol{\kappa} = (\kappa_0, \kappa_1, \ldots, \kappa_L, \mathbf{0}_{P-2L-1}, \kappa_L, \kappa_{L-1}, \ldots, \kappa_1)$$

which are each vectors of length P. The zero padding on the right end of the c_ℓ's is to account for wrap-around effects while the $\boldsymbol{\kappa}$ vector lists the κ_ℓ's in wrap-around order with appropriate zero-padding in the interior. Let $\mathbf{C}$ and $\mathbf{K}$ be the discrete Fourier transforms of $\mathbf{c}$ and $\boldsymbol{\kappa}$ respectively (computed using the FFT) and let $\tilde{\mathbf{F}}$ be the element-wise product of $\mathbf{C}$ and $\mathbf{K}$. The vector $\tilde{\mathbf{f}} = (\tilde{f}_1, \ldots, \tilde{f}_M)$ corresponds to the first M entries of the inverse FFT of $\tilde{\mathbf{F}}$.

EXAMPLE. Suppose that $M = 4$ with grid-counts

$$c_1 = 4, \quad c_2 = 2, \quad c_3 = 1, \quad c_4 = 2.$$

In addition, suppose that the kernel weights are

$$\kappa_0 = 7, \quad \kappa_1 = 5, \quad \kappa_2 = 2$$

so that $L = 2$. While these values are not typical of those arising in practice, their choice ensures simple expressions for illustrative purposes. From (D.2) it is easily shown that the $\tilde{f}_j$ values are

$$\tilde{f}_1 = 40, \quad \tilde{f}_2 = 43, \quad \tilde{f}_3 = 35, \quad \tilde{f}_4 = 23.$$

D.2 THE BINNED KERNEL DENSITY ESTIMATOR

We will now show how these values can be computed using Fourier transform methods. The appropriate value of P is $P = 2^3 = 8$ since it is the lowest power of 2 above $M + L = 6$. The appropriate zero-padded vectors are

$$\mathbf{c} = (4, 2, 1, 2, 0, 0, 0, 0)$$
$$\text{and} \quad \boldsymbol{\kappa} = (7, 5, 2, 0, 0, 0, 2, 5).$$

The Fourier transforms of $\mathbf{c}$ and $\boldsymbol{\kappa}$ can be shown to be

$$\mathbf{C} = (9, 4 - t_1 i, 3, 4 - t_2 i, 1, 4 + t_2 i, 3, 4 + t_1 i)$$
$$\text{and} \quad \mathbf{K} = (21, u_1, 3, u_2, 1, u_2, 3, u_1)$$

where $t_1 = 1 + 2(2^{1/2})$, $t_2 = -1 + 2(2^{1/2})$, $u_1 = 7 + 5(2^{1/2})$ and $u_2 = 7 - 5(2^{1/2})$. The element-wise product of $\mathbf{C}$ and $\mathbf{K}$ is

$$\tilde{\mathbf{F}} = (189, v_1 - w_1 i, 9, v_2 + w_2 i, 1, v_2 - w_2 i, 9, v_1 + w_1 i)$$

where $v_1 = 28 + 20(2^{1/2})$, $v_2 = 28 - 20(2^{1/2})$, $w_1 = 27 + 19(2^{1/2})$, $w_2 = 27 - 19(2^{1/2})$. The inverse Fourier transform of $\tilde{\mathbf{F}}$ is

$$\tilde{\mathbf{f}} = (40, 43, 35, 23, 12, 4, 8, 24).$$

The first $M = 4$ entries of $\tilde{\mathbf{f}}$ correspond to $\tilde{f}_j$, $j = 1, 2, 3, 4$. The remaining entries are polluted by wrap-around effects and should be discarded. ∎

Figure D.2 gives an indication of the accuracy of the binned kernel estimator compared to the exact estimate. In each case the estimates are based on the same data set from the example normal mixture density f_1 as defined by (2.3). The sample size is $n = 1000$ and the bandwidth is $h = 0.22$. The exact estimate was computed over a grid of 1000 points and is shown by the dashed curve. Binned kernel estimates with $M = 15, 25, 50$ and 100 are shown by the solid curves. While there are some obvious discretisation errors for the lower values of M, the binned kernel density estimate based on a value of M as low as 100 is virtually indistinguishable from the exact estimate.

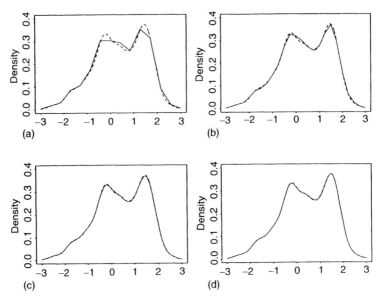

Figure D.2. *Binned kernel density estimates of the normal mixture density f_1 with (a) $M = 15$, (b) $M = 25$, (c) $M = 50$ and (d) $M = 100$. These are shown by solid curves. The dashed curves correspond to the exact kernel density estimate based on 1000 grid points.*

D.3 Computation of kernel functional estimates

As we saw in Chapter 3, a class of functionals that is particularly important in bandwidth selection for kernel density estimation is that with members of the form $\psi_m = \int f^{(m)}(x) f(x)\, dx$ where m is an even integer. As mentioned in Section D.1, direct kernel estimation of ψ_m requires $O(n^2)$ kernel evaluations which makes its computation very expensive for larger sample sizes. In this section we will show how to extend the ideas of the binned kernel density estimator to estimation of ψ_m.

Based on the grid points $g_1, \ldots, g_M$ and corresponding grid counts $c_1, \ldots, c_M$ the binned kernel estimator of ψ_m is

$$\tilde{\psi}_m(h, M) = n^{-2} \sum_{j=1}^{M} \sum_{\ell=1}^{M} K_h^{(m)}(g_j - g_\ell) c_j c_\ell$$

$$= n^{-1} \sum_{j=1}^{M} c_j \left(\sum_{\ell=1-M}^{M-1} c_{j-\ell} \kappa_\ell^{(m)} \right) \quad \text{(D.3)}$$

where

$$\kappa_\ell^{(m)} = n^{-1} K_h^{(m)} \left(\frac{(b-a)\ell}{M-1} \right)$$

for integral values of ℓ. The inner summation in (D.3) is simply a convolution of the c_ℓ and $\kappa_\ell^{(m)}$ and can be computed using the FFT or directly in exactly the same way as was described for the binned kernel density estimator in the previous section. The resulting vector only needs to be multiplied by the vector of grid counts and suitably normalised to obtain $\tilde{\psi}_m(h, M)$.

D.4 Computation of kernel regression estimates

The procedure described in Section D.2 for computation of the binned kernel density estimator can also be easily adapted for fast computation of local linear kernel regression estimates. It is possible to extend the local linear algorithm to polynomials of arbitrary degree but for simplicity we will concentrate on the local linear case here.

It follows from (5.4) that the local linear kernel estimator at a point x has the explicit formula

$$\hat{m}(x; 1, h) = \frac{\hat{s}_2(x; h)\hat{t}_0(x; h) - \hat{s}_1(x; h)\hat{t}_1(x; h)}{\hat{s}_0(x; h)\hat{s}_2(x; h) - \hat{s}_1(x; h)^2}$$

where

$$\hat{s}_r(x; h) = \sum_{i=1}^{n} (X_i - x)^r K_h(X_i - x) \quad \text{and}$$

$$\hat{t}_r(x; h) = \sum_{i=1}^{n} (X_i - x)^r K_h(X_i - x) Y_i.$$

Therefore, it suffices to describe the fast computation of $\hat{s}_r(x; h)$ and $\hat{t}_r(x; h)$.

As in Section D.2, let $g_1, \ldots, g_M$ be a set of grid points on $[a, b]$ with corresponding grid counts $c_1, \ldots, c_M$. The binned approximation to $\hat{s}_r(g_j;h)$ is

$$\tilde{s}_{r,j} \equiv \tilde{s}_r(g_j;h,M) = \sum_{\ell=1}^{M}(g_j - g_\ell)^r K_h(g_j - g_\ell)c_\ell$$

$$= \sum_{\ell=1-M}^{M-1} c_{j-\ell}\kappa_{r,\ell}, \quad j=1,\ldots,M$$

where

$$\kappa_{r,\ell} = \left\{\frac{\ell(b-a)}{M-1}\right\}^r K_h\left(\frac{\ell(b-a)}{M-1}\right).$$

Of course, this is just a straightforward generalization of (D.1) and the $\tilde{s}_{r,j}$ can be computed in the same way as the $\tilde{f}_j$ as described in Section D.2.

Binned approximation of the $\hat{t}_r(g_j;h)$ is similar, although the Y_i's need to be binned first. Let $w_\ell(X_i)$ be the weight assigned to the grid point g_ℓ in the binning scheme applied to the X_i's, so that $c_\ell = \sum_{i=1}^{n} w_\ell(X_i)$. Then the appropriate response variable grid count at g_ℓ is

$$d_\ell = \sum_{i=1}^{n} w_\ell(X_i)Y_i, \quad \ell=1,\ldots,M.$$

The binned kernel approximation to $\tilde{t}_r(g_j;h)$ is therefore

$$\tilde{t}_{r,j} = \sum_{\ell=-L}^{L} d_{j-\ell}\kappa_{r,\ell}, \quad j=1,\ldots,M.$$

The $\tilde{s}_{r,j}$ and $\tilde{t}_{r,j}$ can be combined to give

$$\tilde{m}_{r,j} = \frac{\tilde{s}_{2,j}\tilde{t}_{0,j} - \tilde{s}_{1,j}\tilde{t}_{1,j}}{\tilde{s}_{0,j}\tilde{s}_{2,j} - (\tilde{s}_{1,j})^2}, \quad j=1,\ldots,M,$$

the binned version of $\hat{m}(g_j;h)$.

D.5 Extension to multivariate kernel smoothing

The univariate binned kernel estimators described in the previous sections each have relatively straightforward extensions to their multivariate analogues (Scott, 1992, Wand, 1994). Figure D.3 describes the linear binning strategy for bivariate data. In this case the mass associated with the data point **X** is distributed amongst each of the four surrounding grid points according to areas of the opposite sub-rectangles induced by the position of the data point. The higher-dimensional extension of this rule where areas are replaced by "volumes" is obvious.

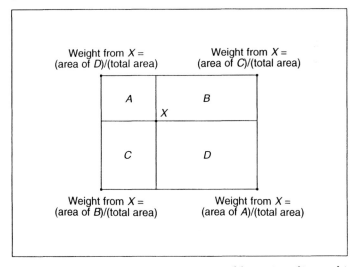

Figure D.3. *Graphical representation of bivariate linear binning.*

In practice there are more stringent constraints on the number of grid points in each direction because of storage and computational costs. In multivariate settings FFT computation of convolutions can lead to considerable savings in computational times although, again, storage restrictions dictate that the amount of zero-padding be kept to a minimum (Wand, 1994).

D.6 Computing practicalities

The binned kernel estimates presented in this appendix are very easy to program. If the FFT is to be used for computation of discrete convolutions then it is useful to have a built-in FFT routine. Such a routine exists in most of the modern statistical programming languages such as GAUSS$_{TM}$ and S-PLUS$_{TM}$. GAUSS$_{TM}$ also has a built-in routine for direct computation of discrete convolutions called CONV. For FORTRAN$_{TM}$ programming, various FFT routines are available through software libraries such as IMSL$_{TM}$ and NAG$_{TM}$.

S-PLUS$_{TM}$ functions for computation of many of the estimators described in this book have been written by the first author and are available in the public domain. Enquiries should be sent to the first author. (His current e-mail address is mwand@hsph.harvard.edu.)

D.7 Bibliographical notes

D.2 Various versions of binned kernel estimators have been proposed and studied by Silverman (1982), Scott and Sheather (1985), Scott (1985) and Härdle and Scott (1992). Härdle and Scott (1992) first showed the full computational advantages of binning. Studies on the accuracy of binned kernel estimators include Hall (1982), Jones and Lotwick (1983), Scott and Sheather (1985), Jones (1989), Härdle and Scott (1992) and Hall and Wand (1994). Further material can be found in Härdle (1990a), Scott (1992) and Fan and Marron (1994).

D.3 For the method presented in this section the authors acknowledge personal communications with Professor J.S. Marron.

D.4 Additional references on binned kernel regression are Georgiev (1986), Härdle (1987), Härdle and Grund (1991) and Fan and Marron (1994).

D.5 Contributions to binned multivariate kernel smoothing include Scott (1992) and Wand (1994).

References

Abramson, I.S. (1982) On bandwidth variation in kernel estimates – a square root law. *Ann. Statist.* **9**, 168–76.

Akaike, H. (1954) An approximation to the density function. *Ann. Inst. Statist. Math.* **6**, 127–32.

Aldershof, B. (1991) Estimation of integrated squared density derivatives. Ph.D. thesis, University of North Carolina, Chapel Hill.

Aldershof, B., Marron, J.S., Park, B.U. and Wand, M.P. (1992) Facts about the Gaussian probability density function. Submitted for publication.

Altman, N. (1992) An introduction to kernel and nearest-neighbor nonparametric regression. *Amer. Statist.* **46**, 175–85.

Bartlett, M.S. (1963) Statistical estimation of density functions. *Sankhyā* Ser. A **25**, 245–54.

Beltrão, K.I. and Bloomfield, P. (1987) Determining the bandwidth of a kernel spectrum estimate. *J. Time Series Anal.* **8**, 21–38.

Bhattacharya, P.K. (1967) Estimation of a probability density function and its derivatives. *Sankhyā* Ser. A **29**, 373–82.

Bhattacharyya, B.B., Franklin, L.A. and Richardson, G.D. (1988) A comparison of nonparametric unweighted and length-biased density estimation of fibres. *Comm. Statist. Theory Meth.* **17**, 3629–44.

Blum, J.R. and Susarla, V. (1980) Maximum deviation theory of density and failure rate function estimates based on censored data. *J. Multivariate Anal.* **5**, 213–22.

Bowman, A.W. (1984) An alternative method of cross-validation for the smoothing of density estimates. *Biometrika* **71**, 353–60.

Bowman, A.W. (1985) A comparative study of some kernel-based nonparametric density estimators. *J. Statist. Comput. Simulation* **21**, 313–27.

Breiman, L. and Friedman, J.H. (1985) Estimating optimal transformations for multiple regression and correlation (with comments). *J. Amer. Statist. Assoc.* **80**, 580–619.

Breiman, L., Meisel, W. and Purcell, E. (1977) Variable kernel estimates of probability density estimates. *Technometrics* **19**, 135–44.

Brillinger, D.R. (1981) *Time Series: Data Analysis and Theory.* Holden-Day, San Francisco.

Brooks, M.M. and Marron, J.S. (1991) Asymptotic optimality of the least-squares cross-validation bandwidth for kernel estimates of intensity functions. *Stochastic Processes Applic.* **38**, 157–65.

Cacoullos, T. (1966) Estimation of a multivariate density. *Ann. Inst. Statist. Math.* **18**, 179–89.

Cao, R., Cuevas, A. and González-Manteiga, W. (1994) A comparative study of several smoothing methods in density estimation. *Comp. Statist. Data Anal.* **17**, 153–76.

Carroll, R.J. and Hall, P. (1988) Optimal rates of convergence for deconvolving a density. *J. Amer. Statist. Assoc.* **83**, 1184–6.

Chiu, S.T. (1991) Bandwidth selection for kernel density estimation. *Ann. Statist.* **19**, 1883–905.

Chiu, S.T. (1992) An automatic bandwidth selector for kernel density estimation. *Biometrika* **79**, 771–82.

Chu, C.K. and Marron, J.S. (1991) Choosing a kernel regression estimator (with discussion). *Statist. Sci.* **6**, 404–36.

Cleveland, W. (1979) Robust locally weighted regression and smoothing scatterplots. *J. Amer. Statist. Assoc.* **74**, 829–36.

Cleveland, W. and Devlin, S. (1988) Locally weighted regression: an approach to regression analysis by local fitting. *J. Amer. Statist. Assoc.* **83**, 596–610.

Cline, D.B.H. (1988) Admissible kernel estimators of a multivariate density. *Ann. Statist.* **16**, 1421–7.

Cline, D.B.H. and Hart, J.D. (1991) Kernel estimation of densities with discontinuities or discontinuous derivatives. *Statist.* **22**, 69–84.

Cox, D.R. (1969) Some sampling problems in technology. In *New Developments in Survey Sampling* (eds. N.L. Johnson and H. Smith). Wiley, New York, pp. 506–27.

Cox, D.R. and Oakes, D.O. (1984) *Analysis of Survival Data.* Chapman and Hall, London.

Davis, K.B. (1975) Mean square error properties of density estimates. *Ann. Statist.* **75**, 1025–30.

Deheuvels, P. (1977a) Estimation non parametrique de la densité par histogrammes generalisés. *Rev. Statist. Appl.* **25**, 5–42.

Deheuvels, P. (1977b) Estimation non parametrique de la densité par histogrammes generalisés (II). *Publ. l'Inst. Statist. l'Univ. Paris* **22**, 1–23.

Devroye, L. (1987) *A Course in Density Estimation.* Birkhäuser, Boston.

Devroye, L. (1989) Consistent deconvolution in density estimation. *Canad. J. Statist.* **7**, 235–9.

Devroye, L. and Györfi, L. (1985) *Nonparametric Density Estimation: The L_1 View.* Wiley, New York.

Diaconis, P. and Freedman, D. (1981) On the histogram as a density estimator: L_2 theory. *Z. Wahr. ver. Geb.* **57**, 453–76.

Diggle, P.J. (1985) A kernel method for smoothing point process data. *Appl. Statist.* **34**, 138–47.

Diggle, P.J. and Marron, J.S. (1988) Equivalence of smoothing parameter selectors in density and intensity estimation. *J. Amer. Statist. Assoc.* **83**, 793–800.

Duan, N. and Li, K.C. (1991) Slicing regression: a link-free regression method. *Ann. Statist.* **19**, 505–30.

Engel, J., Herrmann, E. and Gasser, T. (1995) An iterative bandwidth selector for kernel estimation of densities and their derivatives. *J. Nonparametric Statist.* to appear.

Epanechnikov, V.A. (1969) Non-parametric estimation of a multivariate probability density. *Theory Probab. Appl.* **14**, 153–8.

Eubank, R. (1988) *Spline Smoothing and Nonparametric Regression.* Marcel Dekker, New York.

Fan, J. (1991) On the optimal rates of convergence for nonparametric deconvolution problems. *Ann. Statist.* **19**, 1257–72.

Fan, J. (1992a) Design-adaptive nonparametric regression. *J. Amer. Statist. Assoc.* **87**, 998–1004.

Fan, J. (1992b) Deconvolution with supersmooth distributions. *Canad. J. Statist.* **20**, 155–69.

Fan, J. (1993) Local linear regression smoothers and their minimax efficiencies. *Ann. Statist.* **21**, 196–216.

Fan, J. and Gijbels, I. (1992) Variable bandwidth and local linear regression smoothers. *Ann. Statist.* **20**, 2008–36.

Fan, J., Heckman, N.E. and Wand, M.P. (1995) Local polynomial kernel regression for generalized linear models and quasi-likelihood functions. *J. Amer. Statist. Assoc.* to appear.

Fan, J. and Marron, J.S. (1994) Fast implementations of nonparametric curve estimators. *J. Comput. Graphical Statist.* **3**, 35–56.

Fan, J. and Truong, Y.K. (1993) Nonparametric regression with errors in variables. *Ann. Statist.* **21**, 1900–25.

Faraway, J.J. and Jhun, M. (1990) Bootstrap choice of bandwidth for density estimation. *J. Amer. Statist. Assoc.* **85**, 1119–22.

Farrell, R.H. (1972) On the best obtainable asymptotic rates of convergence in estimation of a density function at a point. *Ann. Math. Statist.* **43**, 170–80.

Fix, E. and Hodges, J.L. (1951) Discriminatory analysis — nonparametric discrimination: consistency properties. *Report No. 4, Project no. 21-29-004*, USAF School of Aviation Medicine, Randolph Field, Texas.

Fix, E. and Hodges, J.L. (1989) Discriminatory analysis — nonparametric discrimination: consistency properties. *Internat. Statist. Rev.* **57**, 238–47.

Földes, A., Rejtö, L. and Winter, B.B. (1981) Strong consistency properties of nonparametric estimators for randomly censored data, II: estimation of density and failure rate. *Periodica Math. Hungar.* **12**, 15–29.

Friedman, J.H. (1991) Multivariate adaptive regression splines (with discussion). *Ann. Statist.* **19**, 1–141.

Friedman, J.H. and Stuetzle, W. (1981) Projection pursuit regression. *J. Amer. Statist. Assoc.* **76**, 817–23.

Friedman, J.H., Stuetzle, W. and Schroeder, A. (1984) Projection pursuit density estimation. *J. Amer. Statist. Assoc.* **84**, 599–608.

Fryer, M.J. (1976) Some errors associated with the nonparametric estimation of density functions. *J. Inst. Math. Appl.* **18**, 371–80.

Fukunaga, K. (1972) *Introduction to Statistical Pattern Recognition.* Academic Press, New York.

Fukunaga, K. and Hostetler, L.D. (1975) The estimation of the gradient of a density function, with applications in pattern recognition. *IEEE Trans. Inf. Theory* **IT-21**, 32–40.

Gasser, T., Kneip, A. and Köhler, W. (1991) A fast and flexible method for automatic smoothing. *J. Amer. Statist. Assoc.* **86**, 643–52.

Gasser, T. and Müller, H.-G. (1979) Kernel estimation of regression functions. In *Smoothing Techniques for Curve Estimation* (eds. T. Gasser and M. Rosenblatt). Springer-Verlag, Heidelberg, pp. 23–68.

Gasser, T. and Müller, H.-G. (1984) Estimating regression functions and their derivatives by the kernel method. *Scand. J. Statist.* **11**, 171–85.

Gasser, T., Müller, H.-G. and Mammitzsch, V. (1985) Kernels for nonparametric curve estimation. *J. Roy. Statist. Soc. Ser. B* **47**, 238–52.

Georgiev, A.A. (1986) A fast algorithm for curve fitting. In *COMPSTAT: Proceedings in Computational Statistics* (eds. F. de Antoni, N. Lauro and A. Rizzi). Physica-Verlag, Vienna, pp. 97–101.

Granovsky, B.L. and Müller, H.-G. (1991) Optimizing kernel methods: a unifying variational principle. *Internat. Statist. Rev.* **59**, 373–88.

Green, P.J. (1987) Penalized likelihood for general semi-parametric regression models. *Internat. Statist. Rev.* **55**, 245–59.

Green, P.J. and Silverman, B.W. (1994) *Nonparametric Regression and Generalized Linear Models.* Chapman and Hall, London.

Grund, B., Hall, P. and Marron, J.S. (1995) Loss and risk in smoothing parameter selection. *J. Nonparametric Statist.* to appear.

Györfi, L., Härdle, W., Sarda, P. and Vieu, P. (1990) *Nonparametric Curve Estimation of Time Series.* Springer-Verlag, Heidelberg.

Hall, P. (1982) The influence of rounding errors on some nonparametric estimators of a density and its derivatives. *SIAM J. Appl. Math.* **42**, 390–9.

Hall, P. (1983) Large sample optimality of least squares cross-validation in density estimation. *Ann. Statist.* **11**, 1156–74.

Hall, P. (1987) On Kullback-Leibler loss and density estimation. *Ann. Statist.* **15**, 1491–519.

Hall, P. (1990) On the bias of variable bandwidth curve estimators. *Biometrika* **77**, 529–36.

Hall, P. (1992) On global properties of variable bandwidth density estimators. *Ann. Statist.* **20**, 762–78.

Hall, P. and Hart, J.D. (1990) Convergence rates in density estimation for data from infinite-order moving average processes. *Probab. Theory Rel. Fields* **87**, 253–74.

Hall, P. and Johnstone, I.M. (1992) Empirical functionals and efficient smoothing parameter selection (with discussion). *J. Roy. Statist. Soc. Ser. B* **54**, 475–530.

Hall, P. and Marron, J.S. (1987a) Extent to which least-squares cross-validation minimises integrated squared error in nonparametric density estimation. *Probab. Theory Rel. Fields* **74**, 567–81.

Hall, P. and Marron, J.S. (1987b) Estimation of integrated squared density derivatives. *Statist. Probab. Lett.* **6**, 109–15.

Hall, P. and Marron, J.S. (1988) Variable window width kernel estimates of probability densities. *Probab. Theory Rel. Fields* **80**, 37–49.

Hall, P. and Marron, J.S. (1991a) Local minima in cross-validation functions. *J. Roy. Statist. Soc.* Ser. B **53** 245–52.

Hall, P. and Marron, J.S. (1991b) Lower bounds for bandwidth selection in density estimation. *Probab. Theory Rel. Fields* **90**, 149–73.

Hall, P., Marron, J.S. and Park, B.U. (1992) Smoothed cross-validation. *Probab. Theory Rel. Fields* **92**, 1–20.

Hall, P., Sheather, S.J., Jones, M.C. and Marron, J.S. (1991) On optimal data-based bandwidth selection in kernel density estimation. *Biometrika* **78**, 263–9.

Hall, P. and Wand, M.P. (1994) On the accuracy of binned kernel density estimators. Submitted for publication.

Härdle, W. (1987) Algorithm AS222. Resistant smoothing using the fast Fourier transform. *Appl. Statist.* **36**, 104–11.

Härdle, W. (1990a) *Smoothing Techniques with Implementation in S.* Springer-Verlag, New York.

Härdle, W. (1990b) *Applied Nonparametric Regression.* Cambridge University Press, Cambridge.

Härdle, W. and Grund, B. (1991) Comment on "Choosing a kernel regression estimator" by C.K. Chu and J.S. Marron. *Statist. Sci.* **6**, 421–5.

Härdle, W., Hall, P. and Marron, J.S. (1988) How far are automatically chosen regression smoothing parameters from their optimum? *J. Amer. Statist. Assoc.* **83**, 86–95.

Härdle, W., Hall, P. and Marron, J.S. (1992) Regression smoothing parameters that are not far from their optimum. *J. Amer. Statist. Assoc.* **87**, 227–33.

Härdle, W. and Marron, J.S. (1985) Optimal bandwidth selection in nonparametric regression function estimation. *Ann. Statist.* **13**, 1465–81.

Härdle, W. and Marron, J.S. (1995) Fast and simple scatterplot smoothing. *Comp. Statist. Data Anal.* to appear.

Härdle, W. and Scott, D.W. (1992) Smoothing in low and high dimensions by weighted averaging using rounded points. *Comput. Statist.* **7**, 97–128.

Härdle, W. and Stoker, T.M. (1989) Investigating smooth multiple regression by the method of average derivatives. *J. Amer. Statist. Assoc.* **84**, 986–95.

Hart, J.D. (1984) Efficiency of a kernel density estimator under an autoregressive dependence model. *J. Amer. Statist. Assoc.*

79, 110–17.
Hart, J.D. and Vieu, P. (1990) Data-driven bandwidth choice for density estimation based on dependent data. *Ann. Statist.* **18**, 873–90.
Hastie, T.J. and Loader, C. (1993) Local regression: automatic kernel carpentry (with discussion). *Statist. Sci.* **8**, 120–43.
Hastie, T.J. and Tibshirani, R.J. (1990) *Generalized Additive Models.* Chapman and Hall, London.
Henderson, H.V. and Searle, S.R. (1979) Vec and vech operators for matrices, with some uses in Jacobians and multivariate statistics. *Canad. J. Statist.* **7**, 65–81.
Hodges, J.L. and Lehmann, E.L. (1956) The efficiency of some nonparametric competitors to the t-test. *Ann. Math. Statist.* **13**, 324–35.
Janssen, P., Marron, J.S., Veraverbeke, N. and Sarle, W. (1995) Scale measures for bandwidth selection. *J. Nonparametric Statist.* to appear.
Jones, M.C. (1989) Discretized and interpolated kernel density estimates. *J. Amer. Statist. Assoc.* **84**, 733–41.
Jones, M.C. (1990) Variable kernel density estimates and variable kernel density estimates. *Austral. J. Statist.* **32**, 361–71.
Jones, M.C. (1991a) The roles of ISE and MISE in density estimation. *Statist. Probab. Lett.* **12**, 51–6.
Jones, M.C. (1991b) Kernel density estimation for length biased data. *Biometrika* **78**, 511–19.
Jones, M.C. (1993) Simple boundary correction for kernel density estimation. *Statist. Computing* **3**, 135–46.
Jones, M.C. (1994) On kernel density derivative estimation. *Comm. Statist. Theory Meth.* **23**, 2133–9.
Jones, M.C., Davies, S.J. and Park, B.U. (1994) Versions of kernel-type regression estimators. *J. Amer. Statist. Assoc.* **89**, to appear.
Jones, M.C. and Foster, P.J. (1993) Generalized jackknifing and higher order kernels. *J. Nonparametric Statist.* **3**, 81–94.
Jones, M.C. and Kappenman, R.F. (1992) On a class of kernel density estimate bandwidth selectors. *Scand. J. Statist.* **19**, 337–50.
Jones, M.C. and Lotwick, H.W. (1983) On the errors involved in computing the empirical characteristic function. *J. Statist. Comput. Simulation* **17**, 133–49.
Jones, M.C., Marron, J.S. and Park, B.U. (1991) A simple root-n bandwidth selector. *Ann. Statist.* **19**, 1919–32.
Jones, M.C., Marron, J.S. and Sheather, S.J. (1992) Progress in

data-based bandwidth selection for kernel density estimation. *Working Paper Series, 92-014*, Australian Graduate School of Management, University of New South Wales.

Jones, M.C., McKay, I.J. and Hu, T.C. (1994) Variable location and scale density estimation. *Ann. Inst. Statist. Math.* **46**, to appear.

Jones, M.C. and Sheather, S.J. (1991) Using non-stochastic terms to advantage in kernel-based estimation of integrated squared density derivatives. *Statist. Probab. Lett.* **11**, 511–14.

Kaplan, E. and Meier, P. (1958) Nonparametric estimation from incomplete observations. *J. Amer. Statist. Assoc.* **53**, 457–81.

Karn, M.N. and Penrose, L.S. (1951) Birth weight and gestation time in relation to maternal age, parity, and infant survival. *Ann. Eugen.* **16**, 147–64.

Kim, W.C., Park, B.U. and Marron, J.S. (1994) Asymptotically best bandwidth selectors in kernel density estimation. *Statist. Probab. Lett.* **19**, 119–27.

Kronmal, R.A. and Tarter, M.E. (1968) The estimation of probability densities and cumulatives by Fourier series methods. *J. Amer. Statist. Assoc.* **63**, 925–52.

Leadbetter, M.R. and Wold, D. (1983) On estimation of point process intensities. In *Contributions to Statistics, Essays in Honor of N.L. Johnson* (ed. P.K. Sen). North-Holland, Amsterdam, pp. 299–312.

Lejeune, M. (1985) Estimation non-paramétrique par noyaux: regression polynomial mobile. *Rev. Stat. Appl.* **33**, 43–67.

Li, K.C. (1991) Sliced inverse regression for dimension reduction (with discussion). *J. Amer. Statist. Assoc.* **86**, 316–42.

Liu, M.C. and Taylor, R.L. (1989) A consistent nonparametric density estimator for the deconvolution problem. *Canad. J. Statist.* **17**, 427–38.

Loftsgaarden, D.O. and Quesenberry, C.P. (1965) A nonparametric estimate of a multivariate density function. *Ann. Math. Statist.* **36**, 1049–51.

Macauley, F.R. (1931) *The Smoothing of Time Series*. National Bureau of Economic Research, New York.

Mack, Y.P. and Müller, H.-G. (1989) Convolution type estimators for nonparametric regression. *Statist. Probab. Lett.* **7**, 229–39.

Mack, Y.P. and Rosenblatt, M. (1979) Multivariate k-nearest neighbor density estimates. *J. Multivariate Anal.* **9**, 1–15.

Magnus, J.R. and Neudecker, H. (1988) *Matrix Differential Calcu-*

lus with Applications in Statistics and Econometrics. Wiley, Chichester.

Mammen, E. (1990) A short note on optimal bandwidth selection for kernel estimates. *Statist. Probab. Lett.* **9**, 23–5.

Marron, J.S. (1989) Comments on a data based bandwidth selector. *Comp. Statist. Data Anal.* **8**, 155–70.

Marron, J.S. (1993) Discussion of: "Practical performance of several data driven bandwidth selectors" by Park and Turlach. *Comput. Statist.* **8**, 17-19.

Marron, J.S. and Nolan, D. (1989) Canonical kernels for density estimation. *Statist. Probab. Lett.* **7**, 195–9.

Marron, J.S and Padgett, W.J. (1987) Asymptotically optimal bandwidth selection for kernel density estimators from randomly right-censored samples. *Ann. Statist.* **15**, 1520–35.

Marron, J.S. and Schmitz, H.P. (1992) Simultaneous density estimation of several income distributions. *Economet. Theory* **8**, 476–88.

Marron, J.S. and Tsybakov, A.B. (1993) Visual error criteria for qualitative smoothing. Unpublished manuscript.

Marron, J.S. and Wand, M.P. (1992) Exact mean integrated squared error. *Ann. Statist.* **20**, 712–36.

McCullagh, P. and Nelder, J.A. (1988) *Generalized Linear Models.* Second Edition. Chapman and Hall, London.

McCune, S.K. and McCune, E.D. (1987) On improving convergence rates for nonnegative kernel failure-rate function estimators. *Statist. Probab. Lett.* **6**, 71–6.

Muirhead, R.J. (1982) *Aspects of Multivariate Statistical Theory.* Wiley, New York.

Müller, H.-G. (1985) Empirical bandwidth choice for nonparametric kernel regression by means of pilot estimators. *Statist. Decisions* Supplement no. 2, 193–206.

Müller, H.-G. (1987) Weighted local regression and kernel methods for nonparametric curve fitting. *J. Amer. Statist. Assoc.* **82**, 231–8.

Müller, H.-G. (1988) *Nonparametric Regression Analysis of Longitudinal Data.* Springer-Verlag, Berlin.

Müller, H.-G. (1991) Smooth optimum kernel estimators near endpoints. *Biometrika* **78**, 521-30.

Muttlak, H.A. and McDonald, L.L. (1990) Ranked set sampling with size-biased probability of selection. *Biometrics* **46**, 435–45.

Nadaraya, E.A. (1964) On estimating regression. *Theory Probab. Appl.* **10**, 186–90.

Nadaraya, E.A. (1974) On the integral mean square error of some nonparametric estimates for the density function. *Theory Probab. Appl.* **19**, 133–41.

Nadaraya, E.A. (Kotz, S. trans.) (1989) *Nonparametric Estimation of Probability Densities and Regression Curves.* Kluwer, Dordrecht.

O'Sullivan, F. and Pawitan, Y. (1993) Multidimensional density estimation by tomography. *J. Roy. Statist. Soc.* Ser. B **55**, 509–21.

O'Sullivan, F., Yandell, B. and Raynor, W. (1986) Automatic smoothing of regression functions in generalized linear models. *J. Amer. Statist. Assoc.* **81**, 96–103.

Padgett, W.J. and McNichols, D.T. (1984) Nonparametric density estimation from censored data. *Comm. Statist. Theory Meth.* **13**, 1581–611.

Park, B.U. and Cho, S. (1991) Estimation of integrated squared spectral density derivatives. *Statist. Probab. Lett.* **12**, 65–72.

Park, B.U. and Marron, J.S. (1990) Comparison of data-driven bandwidth selectors. *J. Amer. Statist. Assoc.* **85**, 66–72.

Park, B.U. and Marron, J.S. (1992) On the use of pilot estimators in bandwidth selection. *J. Nonparametric Statist.* **1**, 231–40.

Park, B.U. and Turlach, B. (1992) Practical performance of several data driven bandwidth selectors (with discussion). *Comput. Statist.* **7**, 251–85.

Parzen, E. (1962) On the estimation of a probability density function and the mode. *Ann. Math. Statist.* **33**, 1065–76.

Patil, P.N., Wells, M.T. and Marron, J.S. (1995) Kernel based estimators of ratio functions. *J. Nonparametric Statist.* to appear.

Prakasa Rao, B.L.S. (1983) *Nonparametric Functional Estimation.* Academic Press, New York.

Press, W.H., Flannery, B.P., Teukolsky, S.A. and Vetterling, W.T. (1988) *Numerical Recipes: The Art of Scientific Computing.* Cambridge University Press, Cambridge.

Priestley, M.B. and Chao, M.T. (1972) Non-parametric function fitting. *J. Roy. Statist. Soc.* Ser. B **34**, 385–92.

Ramlau-Hansen, H. (1983) Smoothing counting process intensities by means of kernel functions. *Ann. Statist.* **11**, 453–66.

Rice, J. and Rosenblatt, M. (1976) Estimation of the log survivor function and hazard function. *Sankhyā* Ser. A **38**, 60–78.

Ripley, B.D. (1994) Neural networks and related methods for classification (with discussion). *J. Roy. Statist. Soc.* Ser.

B **3**, 409-56.

Rosenblatt, M. (1956) Remarks on some nonparametric estimates of a density function. *Ann. Math. Statist.* **27**, 832-7.

Rosenblatt, M. (1991) *Stochastic Curve Estimation*. Institute of Mathematical Statistics, Hayward, CA.

Rudemo, M. (1982) Empirical choice of histograms and kernel density estimators. *Scand. J. Statist.* **9**, 65-78.

Ruppert, D. and Cline, D.B.H. (1994) Transformation kernel density estimation – bias reduction by empirical transformations. *Ann. Statist.* **22**, 185-210.

Ruppert, D., Sheather, S.J. and Wand, M.P. (1995) An effective bandwidth selector for local least squares regression. *J. Amer. Statist. Assoc.* **90**, to appear.

Ruppert, D. and Wand, M.P. (1992) Correcting for kurtosis in density estimation. *Austral. J. Statist.* **34**, 19-29.

Ruppert, D. and Wand, M.P. (1994) Multivariate locally weighted least squares regression. *Ann. Statist.* **22**, to appear.

Sain, S.R., Baggerly, K.A. and Scott, D.W. (1994) Cross-validation of multivariate densities. *J. Amer. Statist. Assoc.* **89**, to appear.

Schucany, W.R. (1989) Locally optimal window widths for kernel density estimation with large samples. *Statist. Probab. Lett.* **7**, 401-5.

Schucany, W.R. and Sommers, J.P. (1977) Improvement of kernel-type density estimators. *J. Amer. Statist. Assoc.* **72**, 420-3.

Scott, D.W. (1979) On optimal and data-based histograms. *Biometrika* **66**, 605-10.

Scott, D.W. (1985) Average shifted histograms: effective nonparametric density estimators in several dimensions. *Ann. Statist.* **13**, 1024-40.

Scott, D.W. (1992) *Multivariate Density Estimation: Theory, Practice, and Visualization*. Wiley, New York.

Scott, D.W. and Factor, L.E. (1981) Monte Carlo study of three data-based nonparametric probability density estimators. *J. Amer. Statist. Assoc.* **76**, 9-15.

Scott, D.W. and Sheather, S.J. (1985) Kernel density estimation with binned data. *Comm. Statist. Theory Meth.* **14**, 1353-9.

Scott, D.W., Tapia, R.A. and Thompson, J.R. (1977). Kernel density estimation revisited. *Nonlinear Anal. Theory Meth. Applic.* **1**, 339-72.

Scott, D.W. and Terrell, G.R. (1987) Biased and unbiased cross-validation in density estimation. *J. Amer. Statist. Assoc.* **82**, 1131-46.

Scott, D.W. and Wand, M.P. (1991) Feasibility of multivariate density estimates. *Biometrika* **78**, 197–206.

Serfling, R.J. (1980) *Approximation Theorems of Mathematical Statistics*. Wiley, New York.

Severini, T.A. and Staniswalis, J.G. (1992) Quasi-likelihood estimation in semiparametric models. Unpublished manuscript.

Sheather, S.J. (1983) A data-based algorithm for choosing the window width when estimating the density at a point. *Comp. Statist. Data Anal.* **1**, 229–38.

Sheather, S.J. (1986) An improved data-based algorithm for choosing the window width when estimating the density at a point. *Comp. Statist. Data Anal.* **4**, 61–5.

Sheather, S.J. (1992) The performance of six popular bandwidth selection methods on some real data sets (with discussion). *Comput. Statist.* **7**, 225–50, 271–81.

Sheather, S.J., Hettmansperger, T.P. and Donald, M.R. (1994) Data-based bandwidth selection for kernel estimates of the integral of $f^2(x)$. *Scand. J. Statist.* to appear.

Sheather, S.J. and Jones, M.C. (1991) A reliable data-based bandwidth selection method for kernel density estimation. *J. Roy. Statist. Soc. Ser. B* **53**, 683–90.

Silverman, B.W. (1978) Density ratios, empirical likelihood and cot death. *Appl. Statist.* **27**, 26–33.

Silverman, B.W. (1982) Kernel density estimation using the fast Fourier transform. *Appl. Statist.* **31**, 93–9.

Silverman, B.W. (1984) Spline smoothing: the equivalent variable kernel method. *Ann. Statist.* **12**, 898–916.

Silverman, B.W. (1986) *Density Estimation for Statistics and Data Analysis*. Chapman and Hall, London.

Silverman, B.W. and Jones, M.C. (1989) E. Fix and J.L. Hodges (1951): an important contribution to nonparametric discriminant analysis and density estimation. *Internat. Statist. Rev.* **57**, 233–8.

Singh, R.S. (1976) Nonparametric estimation of mixed partial derivatives of a multivariate density. *J. Multivariate Anal.* **6**, 111–22.

Singh, R.S. (1979) Mean squared errors of estimates of a density and its derivatives. *Biometrika* **66**, 177–80.

Singh, R.S. (1987) MISE of kernel estimates of a density and its derivatives. *Statist. Probab. Lett.* **5**, 153–9.

Singpurwalla, N.D. and Wong, M.Y. (1983) Kernel estimators of the failure rate function and density estimation: an analogy. *J. Amer. Statist. Assoc.* **78**, 478–81.

Solow, A.R. (1991) An exploratory analysis of the occurrence of explosive volcanism in the Northern Hemisphere, 1851–1985. *J. Amer. Statist. Assoc.* **86**, 49–54.

Staniswalis, J.G. (1989a) Local bandwidth selection for kernel estimates. *J. Amer. Statist. Assoc.* **84**, 284–8.

Staniswalis, J.G. (1989b) The kernel estimate of a regression function in likelihood-based models. *J. Amer. Statist. Assoc.* **84**, 276–83.

Statistical Sciences, Inc. (1991) *S-PLUS Reference Manual*. Statistical Sciences, Inc., Seattle.

Stefanski, L. and Carroll, R.J. (1990) Deconvoluting kernel density estimators. *Statistics* **2**, 169–84.

Stone, C.J. (1977) Consistent nonparametric regression. *Ann. Statist.* **5**, 595–620.

Stone, C.J. (1982) Optimal global rates of convergence of nonparametric regression. *Ann. Statist.* **10**, 1040–53.

Stone, C.J. (1984) An asymptotically optimal window selection rule for kernel density estimates. *Ann. Statist.* **12**, 1285–97.

Swanepoel, J.W.H. (1987) Optimal kernels when estimating non-smooth densities. *Comm. Statist. Theory Meth.* **16**, 1835–48.

Tanner, M.A. and Wong, W.H. (1983) The estimation of the hazard function from randomly censored data by the kernel method. *Ann. Statist.* **11**, 989–93.

Tanner, M.A. and Wong, W.H. (1984) Data-based nonparametric estimation of the hazard function with applications to model diagnostics and exploratory analysis. *J. Amer. Statist. Assoc.* **79**, 174–82.

Tarter, M.E. and Lock, M.D. (1993) *Model-Free Curve Estimation*. Chapman and Hall, London.

Taylor, C.C. (1989) Bootstrap choice of the smoothing parameter in kernel density estimation. *Biometrika* **76**, 705–12.

Terrell, G.R. (1990) The maximal smoothing principle in density estimation. *J. Amer. Statist. Assoc.* **85**, 470–7.

Terrell, G.R. and Scott, D.W. (1985) Oversmoothed density estimates. *J. Amer. Statist. Assoc.* **80**, 209–14.

Terrell, G.R. and Scott, D.W. (1992) Variable kernel density estimation. *Ann. Statist.* **20**, 1236–65.

Tibshirani, R. and Hastie, T. (1987) Local likelihood estimation. *J. Amer. Statist. Assoc.* **82**, 559–68.

Ullah, A. (1985) Specification analysis of econometric models. *J. Quant. Econ.* **2**, 187–209.

Uzunoğullari, U. and Wang, J.L. (1992) A comparison of hazard rate estimators for left truncated and right censored data. *Biometrika* **79**, 297–310.

van Eeden, C. (1987) Mean integrated squared error of kernel estimators when the density and its derivatives are not necessarily continuous. *Ann. Inst. Statist. Math.* **37**, 461–72.

van Vliet, P.K. and Gupta, J.M. (1973) Tham-V – sodium bicarbonate in idiopathic respiratory distress syndrome. *Arch. Dis. Child.* **48**, 249–55.

Vardi, Y. (1982) Nonparametric estimation in the presence of length bias. *Ann. Statist.* **10**, 616–20.

Victor, N. (1976) Nonparametric allocation rules. In *Decision Making and Medical Care: Can Information Science Help?* (eds. F.T. Dombal and F. Grémy). North-Holland, Amsterdam, pp. 515–29.

Wahba, G. (1981) Data-based optimal smoothing of orthogonal series density estimates. *Ann. Statist.* **9**, 146–56.

Wahba, G. (1990) *Spline Models for Observational Data*, SIAM, Philadelphia.

Wand, M.P. (1992a) Error analysis for general multivariate kernel estimators. *J. Nonparametric Statist.* **2**, 1–15.

Wand, M.P. (1992b) Finite sample performance of density estimators under moving average dependence. *Statist. Probab. Lett.* **13**, 109–15.

Wand, M.P. (1994) Fast computation of multivariate kernel estimators. Submitted for publication.

Wand, M.P. and Devroye, L. (1993) How easy is a given density to estimate? *Comp. Statist. Data Anal.* **16**, 311–23.

Wand, M.P. and Jones, M.C. (1993) Comparison of smoothing parameterizations in bivariate kernel density estimation. *J. Amer. Statist. Assoc.* **88**, 520–8.

Wand, M.P. and Jones, M.C. (1994) Multivariate plug-in bandwidth selection. *Comput. Statist.* **9**, 97–117.

Wand, M.P., Marron, J.S. and Ruppert, D. (1991) Transformations in density estimation (with comments). *J. Amer. Statist. Assoc.* **86**, 343–61.

Wand, M.P. and Schucany, W.R. (1990) Gaussian-based kernels. *Canad. J. Statist.* **18**, 197–204.

Watson, G.S. (1964) Smooth regression analysis. *Sankhyā* Ser. A **26**, 101–16.

Watson, G.S. and Leadbetter, M.R. (1963) On the estimation of the probability density, I. *Ann. Math. Statist.* **34**, 480–91.

Watson, G.W. and Leadbetter, M.R. (1964) Hazard analysis I.

Biometrika **51**, 175–84.

Woodroofe, M. (1970) On choosing a delta-sequence. *Ann. Math. Statist.* **41**, 1665–71.

Index

Additive modelling 141, 143
Alternating conditional expectation (ACE) 143
Asymptotic mean integrated squared error (AMISE)
  bandwidth selection 60
  biased cross-validation (BCV) 65
  exercises 88, 89
  plug-in 71
  multivariate kernel density estimation 94–101, 104
  univariate kernel density estimation 21–3
    bibliographical notes 50
    exact 27
    exercises 53, 56
    higher-order kernels 33
    local kernel density estimators 41
Asymptotic mean squared error (AMSE)
  bandwidth selection
    estimation of density functionals 67, 70
    exercises 89
    plug-in 71
  bibliographical notes 50
  local kernel density estimators 41
  regression
    general case 125–6
    locally linear case 120–24
    variable kernel density estimators 43
Asymptotic notation 17–18
Asymptotic relative efficiency (ARE) 106–8, 112
Asymptotic relative mean squared error (ARMSE) 83
Average derivative estimation 143
Average shifted histogram 7

Backfitting 141
Bandwidth
  multivariate kernel density estimation 105–6, 108–9, 110
  regression 138–9, 142
    local polynomial kernel estimators 117
  selection 58–9

biased cross-validation 65–6
bibliographical notes 86–7
comparisons 79–86
estimation of density functionals 67–70
exercises 88–9
least squares cross-validation (LSCV) 63–4
plug-in 71–5
quick and simple 59–62
smoothed cross-validation 75–9
univariate kernel density estimation 12, 13
canonical kernels 29–30
estimation difficulty 36
local kernel density estimators 40, 41
MSE and MISE 16, 19, 20, 22–3
Bandwidth matrix 91, 106–7
Beta density 38
  bandwidth selection 61
  difficulty of estimation 38, 39
  product and spherically symmetric kernels 104
Bias 14, 15
  asymptotic 20–21, 22, 23
  and bandwidth 27
  exercises 52, 56
  higher-order kernels 32, 51
Biased cross-validation (BCV)
  multivariate case 109
  univariate case 65–6, 80–81, 86
Binned kernel density estimator 183–8
Binning 185
Binomial response 164–5
Bins 5–7
Binwidth 5–6
  asymptotic MISE approximation 23, 89
Bivariate kernel density estimation 92–3, 103, 112
  choice of smoothing parametrisation 105–6, 107–8
  random design regression model 115
Biweight kernel 175, 176
  bivariate kernels 103

INDEX

boundary estimation 47–8
  efficiency 31
  exercises 53
  regression 129–30
Blocking method 139
Bootstrap bandwidth selectors 87
Boundaries
  bias 129
  likelihood-based regression 167
  density estimation at 46–9, 52
  kernel regression near 126–30
Box–Cox family 45

Canonical kernels
  multivariate 113
  univariate 28–30
Canonical link function 167
Canonical parameter 167
Cauchy density 175, 176
  error density 171
  exercises 55–6
Censoring variables 154
Characteristic functions 54–5, 175
  errors in data 157
Computation of kernel estimators 182–3
Constant, coefficient 18
CONV 192
Convergence, rate of 18, 23
  bandwidth selectors 70, 80, 81
  higher-order kernels 32, 34
  local kernel density estimators 41
Convolution
  binned kernel density estimators 185–6
  normal probability density function
    multivariate 180
    univariate 178
  univariate kernel density estimation
    14–15
Curse of dimensionality 90, 100, 110, 141

Deconvolution problem 157
Deconvolving kernel density estimator
  157–60, 169, 170–71
Density estimation 5–7, Chapters 2 and 4
  binned 183–8
Density functionals, estimation of 67–70
Dependent data 147–50, 169
Derivative estimation 49, 52
  kernel regression 135–7
Direct plug-in (DPI) rules 71–3, 81, 82
  kernel regression 139
Discrete convolution theorem 186
Discrete Fourier transform 185–6
Discretising data 185
Duplication matrix 96

Effective kernels 133–5, 142
Efficiencies of kernels 31

Epanechnikov kernel 30, 175, 176
  boundary behaviour 127
  effective kernels 134
  efficiency 31, 104–5
  exercises 53
Error density 156
Errors in data measurements 156–60, 169
Exact MISE calculation
  dependent data 149
  multivariate case 101–2, 110
  univariate case 24–7, 50
Exponential density 46
  boundary estimation 48–9
  exercises 55
  likelihood-based regression 166–7
Extreme value densities 175, 176
  difficulty of estimation 39

Failure rate 160
Fast Fourier transform (FFT)
  binned density estimators 183, 184–7
  functional estimators 189
  multivariate kernel smoothing 191
  practicalities 192
Fixed design nonparametric regression 115
  asymptotic MSE approximations 120–22
  comparison of estimators 132
  local polynomial kernel estimators 118
FORTRAN$_{TM}$ 192
Fourier integral density, see Sinc kernel
Functional estimation 188–9

Gamma densities 175, 176
  difficulty of estimation 38, 39
  exercises 56
Gasser–Müller estimator 131, 132, 142
  effective kernels 134, 135
GAUSS$_{TM}$ 192
Generalised linear models 164
Grid counts 183, 185

Hazard function estimation 160–61, 169
Hermite polynomials 177
Heteroscedastic models 115
Histograms 5–7
  inefficiency 23
Hölder's inequality 56
Homoscedastic models 115

IMSL$_{TM}$ 192
Integrated squared density derivatives 67
Integrated squared error (ISE)
  bandwidth selection 80
  univariate kernel density estimation 15
Intensity function estimation 167–8, 170
Inverse discrete Fourier transform 185–6

Kaplan–Meier estimator 155, 156, 169

Kernel estimators 4, 7–8, and throughout
  computation 182–3
Kernel estimators
  binned density 183–8
  functional 188–9
  multivariate smoothing 191
  practicalities 192
  regression 189–90
Kernel spectral density estimator 163

Laplace kernel 175, 176
  errors in data 158, 159
  exercises 55
Leading terms 18
Least squares
  cross-validation (LSCV)
    kernel regression 142
    multivariate case 108–9
    univariate case 63–4, 80–81, 86
  ordinary 1, 2
  weighted 3, 116–17
Leave-one-out density estimator 63, 77
Length biased data 150–54, 169, 170
Likelihood-based regression models 164–7, 170
Linear binning 185
Linear dependence 148
Link function 165
Local kernel density estimators 40–41, 114, 116–19
Local polynomial regression estimates 4
  local linear 3–4, 114–15
    effective kernels 134–5
    exercises 145
Loess 142
Logistic density 175, 176
Logistic regression estimate 165
Lognormal distributions 10–11
  difficulty of estimation 36, 38, 39
  transformation kernel density estimators 44
Long-range dependence 148

Matrices 173–4
Maximal smoothing principle 61
Mean absolute error (MAE) 14
Mean integrated absolute error (MIAE) 16, 50
Mean integrated squared error (MISE)
  bandwidth selection 59, 80
    least squares cross-validation (LSCV) 63
    smoothed cross-validation (SCV) 75, 77
  deconvolving density estimators 159–60
  dependent data 148–50
  multivariate kernel density estimation 101–2

univariate kernel density estimation 15–16
  bibliographical notes 50
  density derivative estimation 49
  exact calculation 24–7, 50
  exercises 52, 53, 55–6
  higher-order kernels 32, 34–5
  see also Asymptotic mean integrated squared error (AMISE)
Mean regression function 155
Mean squared error (MSE)
  bandwidth estimation 89
  univariate kernel density estimation 14–15
    bibliographical notes 50
    density derivative estimation 49
    exercises 57
    higher-order kernels 32
    rate of convergence 18
    variable kernel density estimators 43
  regression 131, 132–3
  see also Asymptotic mean squared error (AMSE)
Moving average dependence 148
Multivariate adaptive regression splines (MARS) 143
Multivariate functions
  normal densities 180–81
  notation 172–3
Multivariate kernel density estimation 90–94
  asymptotic MISE approximations 94–101
  bandwidth selection 108–9
  bibliographical notes 110
  choice of kernel 103–5
  choice of smoothing parametrisation 105–8
  exact MISE calculations 101–2
  exercises 110–13
  nonparametric regression 140–41, 143
Multivariate kernel smoothing 191

Nadaraya–Watson estimator 14, 119, 141
  bandwidth selection 142
  boundary behaviour 129–30
  comparisons 130, 131, 132, 142
  effective kernels 134, 135
  exercises 144, 145, 171
  likelihood-based regression 167, 171
NAG_{TM} 192
Natural parameter 167
Nearest neighbour density estimator 41, 51
Nonparametric regression 3, 4, 10, 114–15
  asymptotic MSE approximations
    general case 125–6
    locally linear case 120–24
  bandwidth selection 138–9
  bibliographical notes 141–3

boundary behaviour 126–30
  comparisons 130–35
  derivative estimation 135–7
  exercises 143–5
  local polynomial kernel estimators
    116–20
  multivariate 140–41
Normal density 175, 176
  difficulty of estimation 39
  efficiency 31
  exercises 55
  multivariate 91–2, 108, 112, 180–81
  univariate 177–80
Normal mixture densities 26
  bandwidth selection 65, 66
  multivariate kernel density estimation
    100, 108, 113
  univariate kernel density estimation 26
  exercises 53
  higher-order kernels 34
Normal scale rules 60, 62, 86

Odd factorial 177
Optimal kernel theory 31
  bibliographical notes 51
Order notation 17, 18
Ordinary least squares line 1, 2
Oversmoothed bandwidth selection rules
  61–2, 86
Oversmoothed estimates 14
Oversmoothing principle 61

Parametric regression 2–3
Parseval's identity 55
Penalised likelihood 170
Periodograms 162–3
Pilot bandwidth
  plug-in bandwidth selection 71
  smoothed cross-validation (SCV)
    bandwidth selection 75, 76, 77–8
Pilot estimation
  local kernel density estimators 41
  variable kernel density estimators 43
Plug-in bandwidth selection
  kernel regression 138–9
    derivative estimation 136, 137
  multivariate case 109
  univariate case 71–5, 81–4, 87
Poisson processes 167–8
Predictor variables 1
Priestly–Chao estimator 130–31, 132
Probability density function 175–6
  estimation 5–7, and throughout
  normal 177–81
Product kernels 91, 103–4
Projection pursuit regression 143

Quick and simple bandwidth selectors 59–62

Radially symmetric kernels 91
Random design nonparametric regression
  115
  asymptotic MSE approximations 123–4
  comparison of estimators 132
  local polynomial kernel estimators
    118–19
Regression 1–2, and throughout
  estimates 4
    computation 189–90
  likelihood-based 164–7, 170
  nonparametric, *see* Nonparametric
    regression
  parametric 2–3
Regression function 115
Relative error 80
Rescaling 28
Response variables 1
Right-censored data 154–6, 169
Root-n bandwidth selectors 84, 87

Scatterplot smoothers 4
Scatterplots 4, 93
Sequences 173
Shifted power family 45
Short-range dependence 148
Simple binning 185
Simulation, bandwidth selectors 85–6
Sinc kernel 35, 51
  characteristic function 55
Sliced inverse regression 143
Smoothed cross-validation (SCV)
  bandwidth selection 75–9, 81–4
  bibliographical notes 87
  stage selection problem 73
Smoothing parameters 6–7, and throughout
Solve-the-equation (STE) rules 74–5, 82, 87
Spectral density estimation 162–3, 169
Spherically symmetric kernels 91, 103–5
Sphering 106, 107–8, 110
S-PLUS$_{TM}$ 192
Stage selection problem 73, 74

Taylor's theorem
  multivariate 94
  univariate 19
Trace 95
Transformation kernel density estimators
  43–5, 51
  exercises 56
Triangular kernel 175, 176
  efficiency 31
Triweight kernel 175, 176
  bandwidth selection 61
  difficulty of estimation 38
  efficiency 3

Ultimately monotone functions 20

Undersmoothed estimates 13
Uniform kernels 175, 176
  efficiency 31, 104
Univariate functions
  normal densities 177–80
  notation 172
Univariate kernel density estimation 10–14
  asymptotic MSE and MISE
    approximations 19–23
  bibliographical notes 50–52
  at boundaries 46–9
  canonical kernels and optimal kernel
    theory 28–31
  derivative 49
  difficulty, measurement of 36–9
  exact MISE calculations 24–7
  exercises 52–7
  higher-order kernels 32–6
  modifications 40–45
  MSE and MISE criteria 14–16
  order and asymptotic notation 17–18
  Taylor expansion 19

Variable kernel density estimators 42–3, 51
Variance
  asymptotic MSE and MISE
    approximation 21, 22
  exercises 52, 53, 56
  mean squared error (MSE) 14
Variance-bias trade-off 22
  bandwidth selectors 66
Variance function 115
Vector-half 95–6
Vectors 95–6
  notation 173–4
Weighted least squares 3, 116–17
Window width, *see* Bandwidth
Wrap-around effects, discrete convolution
  theorem 186, 187

Zero-padding 186, 187